AF412806

Translational Research in Biomedicine

Vol. 6

Series Editor

Samuel H.H. Chan Kaohsiung

Associate Editor

Julie Y.H. Chan Kaohsiung

The Chang Gung Medical Foundation is the patron of this book series.

Shockwave Medicine

Volume Editors

Ching-Jen Wang Kaohsiung
Wolfgang Schaden Vienna
Jih-Yang Ko Kaohsiung

17 figures, 11 in color, 9 tables, 2018

Basel · Freiburg · Paris · London · New York · Chennai · New Delhi ·
Bangkok · Beijing · Shanghai · Tokyo · Kuala Lumpur · Singapore · Sydney

Ching-Jen Wang
Center for Shockwave Medicine and Tissue
Engineering, Department of Orthopedic
Surgery, Kaohsiung Chang Gung Memorial
Hospital and Chang Gung University College of
Medicine
123, Ta-Pei Road
Kaohsiung 83301
Taiwan, ROC

Jih-Yang Ko
Center for Shockwave Medicine and Tissue
Engineering, Department of Orthopedic
Surgery, Kaohsiung Chang Gung Memorial
Hospital and Chang Gung University College of
Medicine
123, Ta-Pei Road
Kaohsiung 83301
Taiwan, ROC

Wolfgang Schaden
Ludwig Boltzmann Institute for Experimental
and Clinical Traumatology
Donaueschingenstraße 13
A–1200 Vienna
Austria

Library of Congress Cataloging-in-Publication Data

Names: Wang, Ching-Jen, 1939- editor. | Schaden, Wolfgang, editor. | Ko,
  Jih-Yang, 1955- editor.
Title: Shockwave medicine / volume editors, Ching-Jen Wang, Wolfgang Schaden,
  Jih-Yang Ko.
Other titles: Translational research in biomedicine ; v. 6. 1662-405X
Description: Basel ; New York : Karger, 2018. | Series: Translational
  research in biomedicine, ISSN 1662-405X ; vol. 6 | Includes
  bibliographical references and indexes.
Identifiers: LCCN 2018001691| ISBN 9783318063127 (hardcover : alk. paper) |
  ISBN 9783318063134 (electronic version)
Subjects: | MESH: Extracorporeal Shockwave Therapy
Classification: LCC RC483.9 | NLM WB 515 | DDC 616.89/122--dc23 LC record available at
  https://lccn.loc.gov/2018001691

© Copyright 2018 by S. Karger AG, P.O. Box, CH–4009 Basel (Switzerland)
www.karger.com
Printed on acid-free and non-aging paper (ISO 9706)
ISSN 1662–405X
e-ISSN 1662–4068
ISBN 978–3–318–06312–7
e-ISBN 978–3–318–06313–4

Contents

Foreword

Welcome to volume 6 of *Translational Research in Biomedicine,* a monograph series dedicated to the dissemination of seminal information in contemporary biomedicine with a translational orientation.

This volume is designed to be a comprehensive reference for shockwave medicine, a relatively new clinical specialty in modern medicine. Originally established as the gold standard for the disintegration of kidney stones, Extracorporeal Shockwave Therapy (ESWT) has progressively evolved to a regenerative treatment modality, and is currently indicated for musculoskeletal disorders and nonskeletal diseases that include ischemic heart disease, diabetic foot ulcers, acute or chronic wounds, and burn lesions. Authored by a panel of international experts and in the spirit of translational medicine, this volume provides a succinct summary of the history and principles of ESWT, the documentations that substantiate the launch of this new clinical specialty, the myriad of physical and cellular or molecular mechanisms that underpin the effects of shockwave, and the future prospects and directions of ESWT.

I wish to express my deep appreciation to Professors Ching-Jen Wang, Wolfgang Schaden, and Jih-Yang Ko, who have taken time from their heavy clinical duties and research endeavors to make this timely volume on "Shockwave Medicine" a reality. I am indebted to the generous patronage of Chang Gung Medical Foundation, Taiwan, which reduces substantially the increasing financial constraints on scientific publication, and allows us to concentrate on publishing timely and crucial themes in translational medicine. I also wish to acknowledge the capable hands of Freddy Brian and Angela Hefti at S. Karger AG during the development and production of this volume. Last but not least, the publication of *Translational Research in Biomedicine* would not have been possible without the foresight, enthusiasm, and whole-hearted support of my dear friend, Dr. Thomas Karger.

Samuel H.H. Chan, Kaohsiung
Series Editor

Preface

Extracorporeal Shockwave Therapy (ESWT) was originally designed for the disintegration of kidney stones and has remained the gold standard for the treatment of urolithiasis because of its clinical success. In orthopedics, an incidental observation of osteoblast responses in bone after ESWT triggered the clinical use of this practice for musculoskeletal disorders. This change resulted in expanding ESWT from a destructive force for stone disintegration into a regenerative treatment modality. Advancements in research and technology further revealed additional mechanisms and pathways of action, and widened the indications of ESWT from musculoskeletal disorders to non-skeletal diseases such as ischemic heart disease, diabetic foot ulcers, acute or chronic wounds, and burn lesions. As such, ESWT has evolved into a multidisciplinary and multifunctional medical subspecialty.

Despite its clinical success, the exact mechanism of ESWT in biological tissue remains unknown. In 1997, Dr. Haupt [1] proposed 4 possible mechanisms for the actions of ESWT on tissue. In its physical phase, ESWT causes a positive pressure to generate absorption, reflection, refraction, and transmission of energy on tissues or cells. Additional studies demonstrated a negative pressure to induce the physical effects such as cavitation and increasing the permeability of cell membrane and ionization of biological tissues. Many signal transductions, including mechanotransduction signal pathway, the extracellular signal-regulated signal kinase, focal adhesion kinase signal pathway, and toll-like receptor 3 signal pathway to regulate gene expression, are activated. In its physical-chemical phase, ESWT stimulates cells to release biomolecules such as ATP to activate signal pathways. ESWT also alters the functions of ion channel in cell membrane and calcium mobilization in cells. In biological tissues, ESWT modulates angiogenesis (vWF, VEGF, eNOS, and PCNA), bone healing (BMP2, osteocalcin, alkaline phosphatase, DKK1 and IGF-1), anti-inflammatory (siCAM and sVCAM) and wound healing (Wnt3 and B-catenin). Furthermore, ESWT stimulates the shift of the macrophage phenotypes from M1 to M2 and increases T-cell proliferation. ESWT activates the toll-like receptor 3 signal pathway to modulate inflammation by controlling the expression of IL6–10 and improves the treatment in ischemic muscles. The future application of ESWT should be combined with advanced and innovative technology, including stem cell therapy, miRNA analysis, gene sequencing, and genomic medicine. This book follows the evolution of ESWT in medicine from the initial stage of destructive ESWT force for lithotripsy to regenerative effects in biological tissues. Section 1 covers the history of shockwave treatment and basic principles. Section 2 includes the application of ESWT in musculoskeletal disorders including osteonecrosis of the femoral head (hip), tendinopathy, fracture treatment, local and systemic effects of ESWT on bone and ESWT and sports related injuries. Section 3 explains the application of ESWT for cardiovascular diseases including preclinical and clinical application of ESWT for ischemic cardiovascular disease and mechanism underlying ESWT for ischemic cardiovascular disease and effects of ESWT on angiogenesis and anti-inflammation-molecular-cellular signaling pathways. Section 4 elaborated on the application of ESWT in urinary disease and other applications including ESWT-assisted intravesical drug delivery and erectile dysfunction and lower urinary tract inflammatory diseases. Section 5 presents the future prospects of shockwave medicine including current application and future procedures of ESWT.

We are grateful to Karger Publishers and to the Series Editor, Professor Samuel H.H. Chan for the opportunity to publish this volume. This book commemorates the two-year anniversary of the Center for Shockwave Medicine and Tissue Engineering at Kaohsiung Chang Gung Memorial Hospital. This book contains the most complete and up-to-date ESWT-related data and documents with worldwide implications. We are committed to pursue ESWT-related research and new discovery with innovative procedures along with the new indications in shockwave medicine.

Ching-Jen Wang, Kaohsiung
Wolfgang Schaden, Vienna
Jih-Yang Ko, Kaohsiung

Reference

1　Haupt G: Use of extracorporeal shock waves in the treatment of pseudarthrosis, tendinopathy and other orthopedic diseases. J Urol 1997;158:4–11.

Wang C-J, Schaden W, Ko J-Y (eds): Shockwave Medicine.
Transl Res Biomed. Basel, Karger, 2018, vol 6, pp 1–16 (DOI: 10.1159/000485050)

History of Shockwave Treatment and Its Basic Principles

Kenneth Craig S. Vincent[a, b] · Maria Cristina d'Agostino[c]

[a]Kompass-FlashWave Regenerative Centre, Auckland, New Zealand; [b]Kompass-FlashWave Regenerative Centre, Victoria, Australia; [c]Shock Wave Therapy and Research Unit, Humanitas Clinical and Research Hospital, Milan, Italy

Abstract

Shockwaves, which have had several geophysical applications, were successfully introduced into the field of medicine over 3 decades ago for the purpose of eradicating urolithiasis, heralding the era of noninvasive medical intervention. Technological advancements in the area of medical shockwaves made in recent years have broadened the spectrum of its clinical application from being just purely a destructive force into a treatment modality that engenders a myriad of progenesis effects associated with tissue regeneration and functional restoration. Although the exact mechanisms of action *(stimulodynamics)* of medical shockwaves are yet to be completely elucidated, observation of its effect on tissue revealed; bio-chemical and biocellular modulation, resulting in progenesis effects such as; angiogenesis, osteogenesis, and tendogenesis. These progenesis responses are considered to be precipitated via a sensory derived signal transduction effect *(stimulokinetics)*, commonly known as mechanotransduction. Three decades later medical shockwaves remains a phenomenon that is intransigent against becoming demoded, and continues to see its clinical utilty expand across medical disciplines. This expansion highlights the need for adequate training and education. The safety, systemic neutrality, noninvasive nature, and regenerative properties associated with medical shockwave treatment warrants further research and deserves greater recognition in the global healthcare system. The expansion of the clinical utility of medical shockwaves could elevate the management of multiple pathologies from being merely palliative, toward a more sanative approach, improving the quality of life of patients across their lifespan, while reducing the burden of healthcare costs globally.

Introduction

The theory for extracorporeal calculi disintegration utilizing pressure waves has been explored since the 1950s, initially utilizing ultrasound waves [1–3], but its clinical viability was hampered due to the excessive tissue damage associated with this procedure. The geophysical observation of shockwave (SW) propagation by rapid-velocity raindrops, and micrometeorites by aerospace engineers, stimulated much interest and research of this phenomenon. The impact of SWs on human tissue was originally noted during World War II where underwater explosions due to depth charges caused internal tissue damage to the lungs of castaways devoid of any appreciable external evidence of trauma or violence. Professor Eberhard Häusler's collaboration with physicians from the University of Munich and technicians from Dornier pioneered the concept and investigation of kidney stone disintegration by extracorporeal SWs [4, 5]. In 1974, a research grant titled "Application of Shockwave Lithotripsy" was awarded, and in 1980, the first patient was treated with extracorporeal shockwave lithotripsy (ESWL) [5–7]. This pioneered and revolutionized minimally invasive intervention in urology [5–7], and to date, ESWL remains the gold standard for the treatment of urolithiasis. The use of SWs was first introduced into orthopedics for the emancipation of bone cement removal in the late 1980s [7–11], but ancillary observations in 1991 of the osteoblastic responses to SW on bone [5, 8–10] would change the utility of SWs from being merely a destructive force for the disintegration and emancipation into a regenerative treatment modality. Advancements in technology have witnessed the expansion and utility of SWs in orthopedics [7, 8, 12–22], musculoskeletal medicine [23–48], vulnology [49–52], andrology [53–59], and more recently interventions in cardiology [60–66], spinal cord injury [67–70], interventions in ageing (sarcopenia), and musculoskeletal tissue resilience [71, 72]. Dose dependent stimulus from SWs are seen to engender tissue regeneration and restoration in various pathologies and more investigations are being undertaken to obtain greater elucidation about the transmission (stimulokinetics) and mechanisms of action (stimulodynamics) of medical shockwave treatment (SWT) on tissue.

SW: Basic Characteristics and Methods of Propagation

It is pertinent to note that an SW differs from an ultrasound wave, in that SWs are biphasic supersonic waves that transmit in a non-sinusoidal motion pattern devoid of thermal and micro lesion effects, and achieve peak pressure amplitudes over a thousand times greater than that of an ultrasound wave [26, 28, 34, 76, 78, 79]. In 1997, an International Consensus Conference (1997) established the physical characteristics of an SW (Fig. 1).

Shockwaves utilized in medicine bear these characteristics (Fig. 1) and are propagated utilizing electrohydraulic (Fig. 2a), electromagnetic (Fig. 2b), or piezoelectric

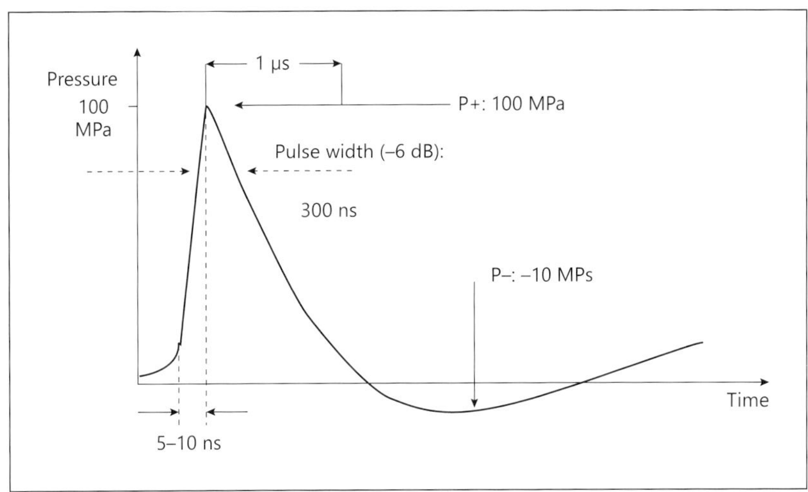

Fig. 1. Characteristics of a shockwave: phase 1: high pressure wave rise-time from basic ambient value, to a pressure value of approximately 100 MPa within <10 nanoseconds (ns). Phase 2: wave implosion to a negative pressure value of approximately –10 MPa within microseconds. Image adapted from [8].

(Fig. 2c) technology [35, 76, 78, 79]. SWs created by each of these 3 sources are primarily propagated by a contained high-voltage discharge within a three-dimensional (3D) fluid-filled chamber that causes an ephemeral high pressure disturbance, where each discharge propagates a biphasic sonic impulse (Fig. 1) within the chamber. The accelerated and sudden rise from ambient pressure within the chamber creates an extremely short duration broad frequency spectrum (16–20 MHz) SW, that rises to its peak pressure (100 MPa) and implodes (–10 MPa) within nanoseconds (Fig. 1) of its lifecycle [26, 28, 35, 76, 78, 79].

In order to ensure minimal attenuation of the shockwave's energy at the refraction point (entry point onto target tissue), ultrasonic gel is utilized for maximal force transmission. Modern SW devices are capable of producing both focused and unfocused SW impulses of varying penetration depths and energy flux densities in order to cater to multiple treatment parameters. Pertinent factors to consider when comparing SWT technology are; pressure distribution, focal zone area, energy flux density, and the total energy concentration at the second wave (focal refraction) zone [28, 29, 34]. More recently the introduction of radial pulse devices have emerged, and due to the lower economic cost associated with radial type devices, its use has increased in popularity. It is of great importance to note that radial pulse devices often referred to as "radial shockwave therapy" produce a wave that has completely divergent physical characteristics (Fig. 3) from those of medical shockwaves as classified by the International Consensus Conference 1997 and as described in Figure 1 [28, 78, 79, 81, 84, 85]. Radial pulses are propagated by either compressed air or by a magnetic motor, where a metallic projectile within a barrel chute in the applicator

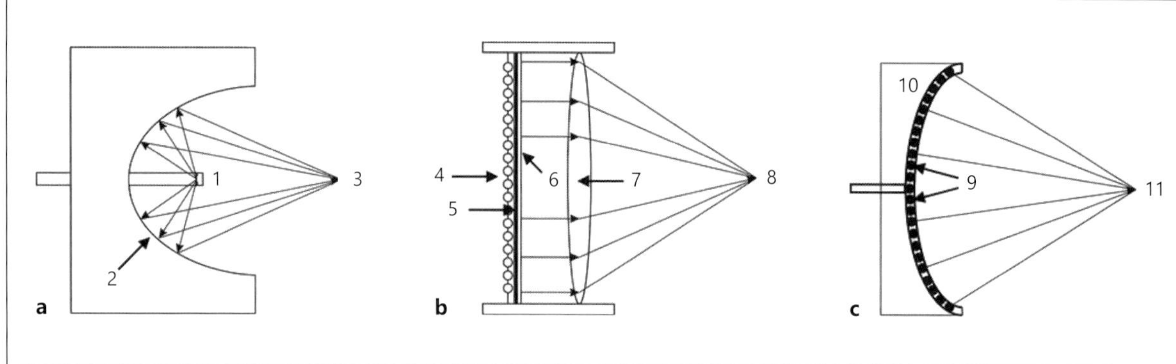

Fig. 2. a Shockwave (SW) generated by an electrohydraulic device. At the focal point (1), an electrical discharge is released by an electrode causing heat water vaporization within the chamber. This generates a gas bubble to be rapidly filled with water vapor and plasma. The result of this extremely rapid expansion of the bubble is a sonic pulse, and the subsequent implosion of this bubble causing a reverse pulse, creating an SW. Wave reflection occurs with the assistance of a reflector (2), and the SW is converted and propelled forward as an acoustic pressure pulse. The point of highest pressure occurs at the secondary wave region (3). The secondary wave region or the wave refraction point is where the SW is aimed and transmitted into the desired treatment region. Image adapted from [8, 28, 35, 76]. **b** SW generated by an electromagnetic device. An adjustable magnetic field is generated by passing potent electric current via a coil (4), causing a high current in an antithetical metal membrane (5). The surrounding liquid in the adjacent membrane (6) is forced rapidly away (reflection site). Due to the high conductivity of the adjacent membrane, the liquid is forced away rapidly, and the ensuing compression of the surrounding liquid generates a SW. An acoustic lens (7) focuses the SW and transmits the wave to the secondary wave region (8). The secondary wave region or the wave refraction point is where the SW is aimed and transmitted into the desired treatment region. Image adapted from [8, 28, 35, 76]. **c** SW generated by an piezoelectric device. Several hundred ceramic piezocrystal transducers (9) are arranged in a mosaic pattern on a bowl-shaped carrier at the reflection site (10). Upon a rapid electrical discharge, the piezocrystals react with an expensory deformation (inverse piezoelectric effect), that propagates a pressure pulse in the surrounding fluid and is directly self-focused at the secondary focal region (11). The secondary wave region or the wave refraction point is where the SW is aimed at and transmitted into the desired treatment region. Note: For latest updates on the various technologies, readers should contact each manufacturer directly for propriety infromation. Image adapted from [8, 28, 35, 76].

is rapidly accelerated linearly within the chute. The ensuing ballistic energy occurring at the tip of the applicator (refraction point) is placed onto the skin of the target region, and this energy is then transferred onto the target tissue region as spherical radial pulse waves [28, 78, 79, 81, 84, 85]. The following are the key factors of wave divergence between medical shockwave treatment (SWT), and radial pulse therapy (RPT): principle of stimulus propagation, wave length, maximal energy pressure, wave speed, penetration depth, focal zone size, and maximal energy at the secondary focal (refraction) region. Although SWs, ultrasound waves, and radial pulses are considered as being acoustic waves, they each have completely divergent character-

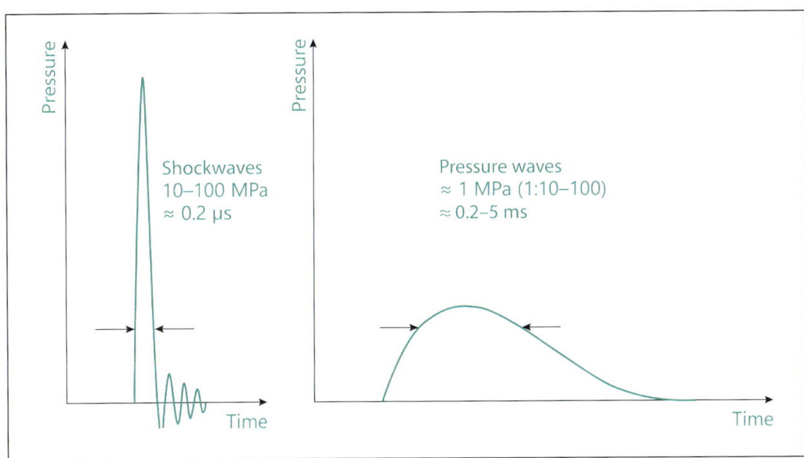

Fig. 3. Characteristic divergence between a shockwave (SW; left), and a radial pressure pulse (right). Radial waves do implode to a negative pressure value (Figure 4), and is not depicted in this illustration. The negative pressure of radial waves occurs slower when compared to SWs. Image adapted from [84, 85].

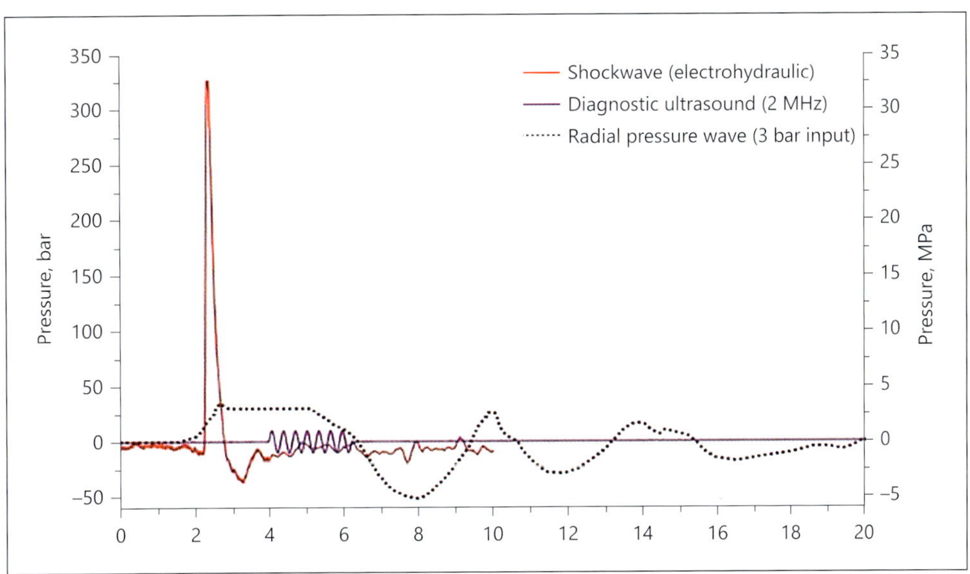

Fig. 4. Comparative characteristic divergence between an SW (Solid Red), ultrasound wave (Solid Blue), and a radial pulse wave (Dotted Black). Wave pattern, peak pressure, speed of wave rise time, and implosion. Image adapted from [85].

istics (Fig. 4), and will each have a unique action, influence on tissue, and clinical outcome.

SWT is developing exponentially along with advances in technology (Fig. 5a–d), and the need for appropriate professional education and training becomes essential in order to obtain optimal clinical outcomes while ensuring patient safety. When in-

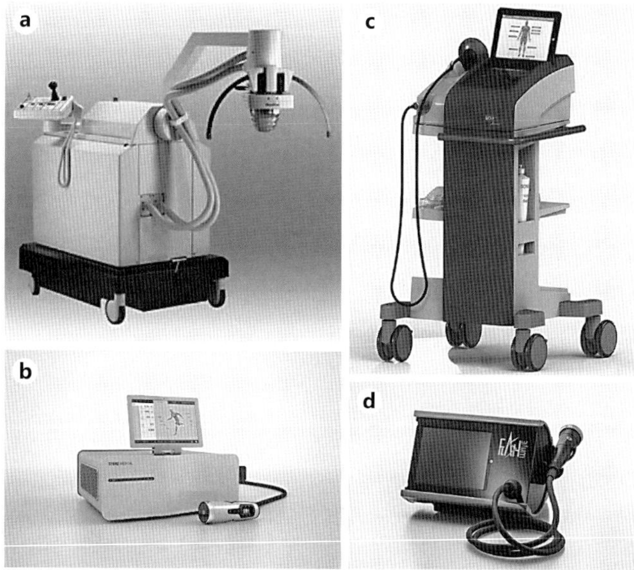

Fig. 5. a The 1st-generation orthopedic extracorporeal shockwave (SW) device: Ossatron (electrohydraulic). Manufacturer: High Medical Technologies. This device was a strictly focused SW device. **b** Newer generation SW device: Duolith SD-1 Ultra® (electro-magnetic). Manufacturer: Storz Medical AG. **c** Newer generation multiple application extracorporeal SW device: RWPiezowave2® (Piezoelectric). Manufacturer: Richard Wolf GmbH/Elvation Medical GmbH. **d** Latest evolution FlashWave™ technology (electrohydraulic) Manufacturer: NonVasiv GmbH.

correct treatment protocols (i.e. energy density flux level, treatment applicator placement, etc.) and technology are utilized, treatment outcomes are often severely compromised and often lead to poor clinical results [41, 86, 87]. When medical SWT is performed by an experienced and licensed clinician with the appropriate selection of technology, it has proven to yield excellent and sustainable clinical outcomes across a broad spectrum of pathologies [7, 8, 12–70, 77–83] . Therefore, the importance is underscored to obtain adequate professional training and education, as it is a key factor for clinical success, and patient safety.

The European Society for Musculoskeletal Shockwave therapy (ESMST) was founded in 1997. The ESMST and its founding members actively conducted research and established treatment guidelines and protocols for SWT. In 1999, the ESMST was renamed as the International Society for Medical Shockwave Treatment (ISMST) due to the growing number of international participation and interest in ESWT. The ISMST is the society presently responsible for research, training, and education for medical shockwave treatment, and RPT. The ISMST and its affiliated regional societies hold regular scientific congresses and certification courses (www. ismst.com)

SWT Stimulodynamics: Mechanisms of Action and Biological Responses

Pathophysiologies especially in chronic states are multifactorial and are often indocile to most treatments and remain an enigma for clinicians to ameliorate. Dissimilar from its use for the eradication of urolithiasis at its inception, the utility of medical shockwave treatment (SWT) today is considered for its regenerative properties to address impervious conditions encountered by various medical disciplines as mentioned earlier in the chapter. *Stimulodynamics* may be considered as the action or influence exacted by a particular force on tissue, and the ensuing biochemical and biocellular responses derived from it. Although the exact mechanisms of action (stimulokinetics) of SWs on tissue is yet to be completely elucidated (as is the case for most if not all medical interventions), researchers (i.e., Schaden et al. [13], Ogden et al. [26], Wang [28], Notanicola and Moretti [35], Mittermayr et al. [51], among others) have provided insights into the biological responses associated with SWT and the regenerative outcomes observed in non-union fractures, tendinopathies, and chronic wounds [13, 21, 26, 35, 51]. More recently, ground-breaking investigations by Lobenwein et al. [69], Sukubo et al. [89], and Holfeld et al. [90] have provided greater elucidation into the possible mechanics of action of SWT on tissue. These investigations revealed that a dose-dependent signal transduction from SWT influenced toll-like receptors, macrophages, and interleukins leading to a cascade of biochemical and biocellular responses commencing with angiogenesis [65, 69, 87–90], which then translates into a combination of regenerative processes leading toward homeostatic return and function. The culminating biochemical and biocellular responses influenced by SWT are seen to effectively address multiple aberrances associated with complexities, which include *modulation of neurotransmitters* (i.e., Substance P Calcitonin Gene-Related Peptide), and inflammatory factors (i.e., prostaglandin, interleukins), encouraging analgesia and pain modulation in chronic and even complex pain conditions [41, 91–93, 131]; *Progenitor cell, mescenchymal stem cell,* and *growth factor* proliferation (i.e., vascular endothelial growth factor, transforming growth factor β-1) [28, 35, 94–99] are all elements seen to assist with progenesis responses such as angiogenesis [49–70, 88–90, 97–99], osteogenesis [8, 12–22, 100–103], and tendogenesis [28–45, 94, 95].

Stimulokinetics of SWT

Stimulokinetics provide insights as to how forces (shear, tension, and compression) are detected and transduced by cells, which is then translated into regulatory signals. Any structure regardless of being inanimate or animate require the actions of physical forces on them. The constant attraction-repulsion force feedback loop acting on its atoms, allow all structures to hold their form, and perform their respective functions. Cells are considered to be the most essential and fundamental components of all living organisms, and were first closely investigated by Robert Hooke in the late 1600s. The principles of physics described in *"Hooke's Law"* regarding the influence of force on matter would be later adapted into active force influences in cellular biology by Wilhelm His in the late 1800s. The detection and transcription of forces into biochemical signals are key fundamentals by which cells and organs maintain viability, stability and health [115–124]. The stimulokinetics and the ensuing translational responses of cells to force (stimulodynamics) are commonly termed as *"mechanotransduction."* Over the past two decades, resurgence in the study of mechanobiology has provided greater insights of cellular responses to stimulus (forces), and its role in homeostasis and disease [115–124]. To be concise, mechanotransduction involves the communication and modulation of forces on molecular and cellular levels determining the cellular and organism's fate. The cytoskeleton structure of mammalian cells are able to detect and translate forces in minute nano-Newton ranges, and play a pertinent role in homeostasis (i.e., stem cell differentiation) and in disease (i.e., neoplastic processes) [125–128]. Protein folding, bone shaping, muscle contraction and regeneration, hearing, touch, lung regulation, and circulatory pressure are all in principle regulated by transducers that sense, respond, or react to this mechanosensory feedback complex [115–124]. Although the exact mechanism and the complexities of mechanotransduction is yet to be fully elucidated, chemical signals and tensional translations of mechanical forces are considered as being a key precipitator of events [115–131] – example, fluid force (shear stress) influences arterial blood, traction force (tensional stress) influences cell migration, and membrane tension (tensional stress) influences muscle contraction channels, all with a unifying principle that these force-translation exchanges are linked to the extracellular matrix and the cytoskeleton by a network of specialized transducers [129–131] and receptor sites. Medical SWT is considered to exert its influence on tissue by means of stimulus translation via this mechanotransduction principle [77]. The high velocity supersonic force generated during SWT is transmitted into the target tissue where a nonviolent isotropic shear stress force occurs on cell membranes leading to microvesicle excuviations. The excuviated microvesicles carry potent cargo (i.e., proteins, nucleic acids) that stimulates neighboring cells by fusion, which in turn activates cellular communication and the biochemical and biocellular translational effects (Fig. 6).

Simultaneously these very same SW impulses exert a pressure spectrum that causes a contractile tensional force that acts on the integrin, and actin cytoskeleton complex of cells. Hence, the stimulus from SWT is seen to exert its influence on both the extra and intracellular matrix [77]. The translational effects of SWT on tissue are seen to influence a broad spectrum of endogenous biochemical and biological events (i.e., toll-like receptors, inflammatory markers, neurotransmitters, and progenesis factors) that have been known to engender homeostatic return and functional restoration [7, 8, 12–27, 35, 40, 41, 49, 50–59, 62–67, 88–103]. From what is presently known of the microenvironment (protein network and structural matrix composition) of cells, it becomes plausible to hypothesize the mechanotransduction principle of SWT force action on tissue, providing some clarity as to the stimulokinetics and stimulodynamics of SWT on cells. It is further plausible when considering the mechanics of action and the biological progenesis responses associated with SWs, to

Vincent · d'Agostino

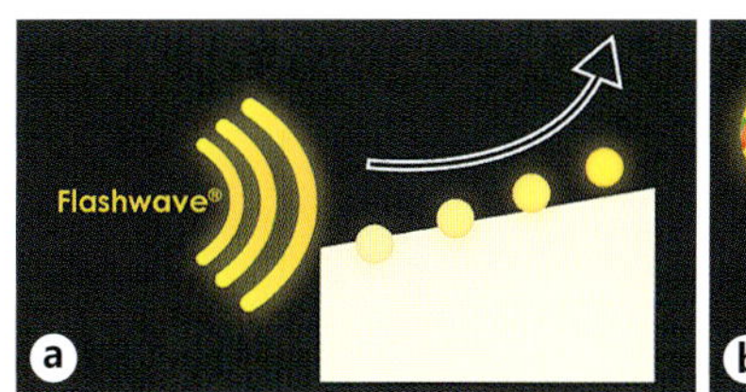

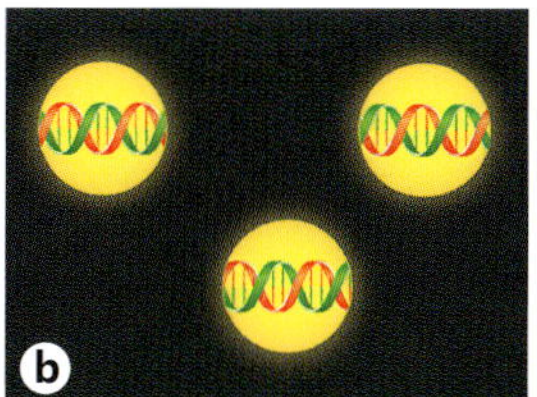
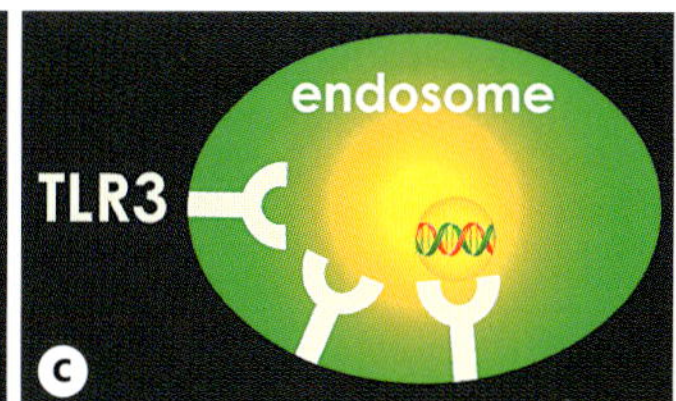

Fig. 6. Kinetic action of SWT on cells [77]. Shearing force from SWT (**a**) causes excuviation and excursion of cargo rich microvesicles (**b**), ensued by endosomal activation of toll-like receptors triggering modulation of macrophages, interleukins etc (**c**). These proceses are seen to engender progenesis effects (ie. angiogenesis), and the ensuing cascade of metabolic activities involving tissue regeneration [28, 35, 49–70, 87–90, 94–99]. Note: Image with permission of NonVasiv GmbH.

consider medical SWT as being an endogenous stem cell activation treatment modality [71, 72, 104–114, 133, 134] when the appropriate technique and technology is selected and applied.

The Future of SW Medicine

We can usually tell the future by examining the past; over the past 3 decades, the history of medical SWT has revealed that its safety and efficacy in each area of its utility has encouraged further research for its expansion, while remaining intransigent to becoming demoded. There is growing evidence that SWT is influencing the activation and differentiation of endogenous stem cells, and is stimulating more research in this area. Among such investigations are: the utility of the evolution FlashWave™ technology for musculogenesis via endogenous skeletal muscle stem cell activation and differentiation, presently being investigated by Vincent et al. [71, 72] to determine the efficacy for skeletal muscle regeneration in the aging population (sarcopenia/dynapenia), and to improved skeletal muscle resilience in the athletic population. Holfeld et al. [65] are presently investigating intracorporeal SWT on the heart (Fig. 7a, b) ushering in an interesting era in SW medicine.

In urology where SW were first utilized (ESWL) as a destructive force, SWT is now utilized as a regenerative and restorative modality redefining andrology, addressing the issue of erectile dysfunction and is showing very promising results [53–59]. Other investigations include the utility of SWT in dystonia and spasticity [135–140], diabetic neuropathy [141, 142], knee osteoarthritis [143–145], low back pain syndromes [146], uterine fibroids [147], and possible intervention in Langerhans islet beta-cell regeneration in type-1 diabetes mellitus [148], among other emerging indications. The potential clinical utility of SWT is vast and more vigorous research, clinical exploration and collaboration is required to provide firm support for its use and to receive the recognition and funding from both the public and private healthcare sectors.

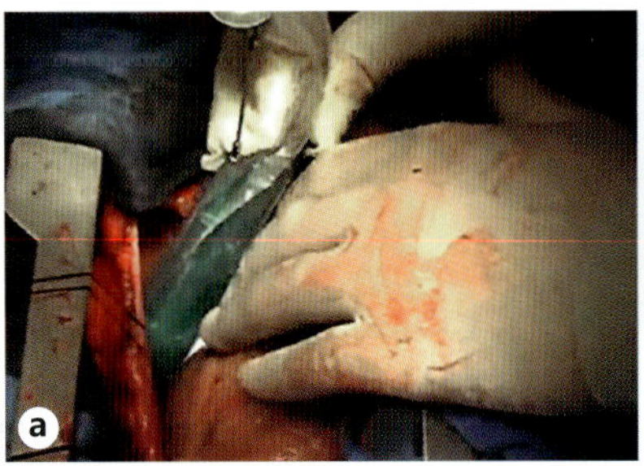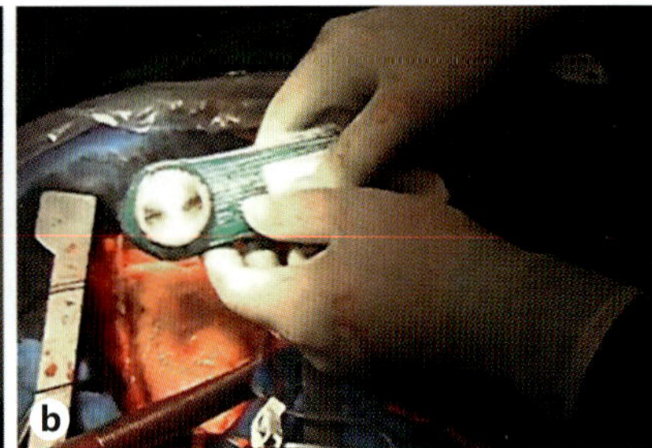

Fig. 7. a Intracorporeal shockwave (SW) treatment of the heart: Holfeld et al. [64], Dept. of Cardiac Surgery, Innsbruck Medical University, Austria. **b** Specialised intracorporeal SW (electrohydraulic) applicator designed for open heart interventions.

Conclusion

With over 3 decades of utility since its inception into medicine, SWT has continued to extend its utility over multiple disciplines and has continued to be widely investigated by researchers globally. No two devices will produce the exact same stimulus, and this fact will have an influence and impact on clinical efficacy and outcomes. Therefore, the selection of appropriate technology, and the employment of effective assessment and application techniques is a pertinent fact for clinical success. The safety, noninvasive nature, and systemic neutrality of SWT makes it a unique interventional modality, and the expansion of its utility may provide a more sanative approach for the management of multiple pathologies. The increased utility of SWT has the potential to improve the quality of life and productivity across the lifespan, while reducing the burdens of global healthcare costs.

References

1 Lamport H, Newman HF, Eichhorn RD: Fragmentation of biliary calculi by ultrasound. Fed Probat 1950; 9:73–74.

2 Coats EC: The application of ultrasonic energy to urinary and biliary calculi. J Urol 1956;75:865–874.

3 Mulvaney WP: Attempted disintegration of calculi by ultrasonic vibrations. J Urol 1953;70:704–707.

4 Wess O: Physics and technique of shock wave lithotripsy (SWL); in Talai J, Tieselius HG, Albala DM, Ye Z (eds): Urolithiasis. Basic Science and Clinical Practice. London, Springer Verlag, 2012, pp 301–311.

5 Loske AM: Brief historical background; in Graham RA, Davidson L, Horie Y (eds): Medical and Biomedical Applications of Shock Waves. Springer International Publishing AG, Switzerland, 2017, pp 7–8.

6 Herr HW: "Crushing the stone:" a brief history of lithotripsy, the first minimally invasive surgery. BJU Int 2008;102:432–435.

7 Thiel M: Application of shock waves in medicine. Clin Orthop Relat Res 2001;387:18–21.

8 Ueberle F: Shock wave technology; in Siebert W, Buch M (eds): Extracorporeal shock waves in orthopaedics. Berlin, Springer, 1997, pp 59–87.

9 Karpman RR, Magee FP, Gruen TW, Mobley T: The lithotriptor and its potential use in the revision of total hip arthroplasty. Orthop Rev 1987;16:81–85.

10 Weinstein JN, Oster DM, Park JB, Park SH, Loening S: The effect of the extracorporeal shock wave lithotriptor on the bone-cement interface in dogs. Clin Orthop Relat Res 1988;235:261–267.

Vincent · d'Agostino

11 May TC, Krause WR, Preslar AJ, Smith MJ, Beaudoin AJ, Cardea JA: Use of high-energy shock waves for bone cement removal. J Arthroplasty 1990;5:19–27.

12 Valchanou VD, Michailov P: High energy shock waves in the treatment of delayed and nonunion of fractures. Int Orthop 1991;15:181–184.

13 Schaden W, Fischer A, Sailler A: Extracorporeal shock wave therapy of nonunion or delayed osseous union. Clin Orthop Relat Res 2001;387:90–94.

14 Alves EM, Angrisani AT, Santiago MB: The use of extracorporeal shock waves in the treatment of osteonecrosis of the femoral head: a systematic review. Clin Rheumatol 2009;28:1247–1251.

15 Cacchio A, Giordano L, Colafarina O, Rompe JD, Tavernese E, Ioppolo F, Flamini S, Spacca G, Santilli V: Extracorporeal shock-wave therapy compared with surgery for hypertrophic long-bone nonunions. J Bone Joint Surg Am 2009;91:2589–2597.

16 Furia JP, Juliano PJ, Wade AM, Schaden W, Mittermayr R: Shock wave therapy compared with intramedullary screw fixation for nonunion of proximal fifth metatarsal metaphyseal-diaphyseal fractures. J Bone Joint Surg Am 2010;92:846–854.

17 Furia JP, Rompe JD, Cacchio A, Maffulli N: Shock wave therapy as a treatment of nonunions, avascular necrosis, and delayed healing of stress fractures. Foot Ankle Clin 2010;15:651–662.

18 d'Agostino C, Romeo P, Amelio E, Sansone V: Effectiveness of ESWT in the treatment of Kienböck's disease. Ultrasound Med Biol 2011;37:1452–1456.

19 Vulpiani MC, Vetrano M, Trischitta D, Scarcello L, Chizzi F, Argento G, Saraceni VM, Maffulli N, Ferretti A: Extracorporeal shock wave therapy in early osteonecrosis of the femoral head: prospective clinical study with long-term follow-up. Arch Orthop Trauma Surg 2012;132:499–508.

20 Leal C, d'Agostino C, Gomez Garcia S, Fernandez A: Current concepts of shockwave therapy in stress fractures. Int J Surg 2015;24:195–200.

21 Schaden W, Mittermayr R, Haffner N, Smolen D, Gerdesmeyer L, Wang CJ: Extracorporeal shock-wave therapy (ESWT) – first choice treatment of fracture non-unions? Int J Surg 2015;24(pt B):179–183.

22 D Agostino MC, Frairia R, Romeo P, Amelio E, Berta L, Bosco V, Gigliotti S, Guerra C, Messina S, Messuri L, Moretti B, Notarnicola A, Maccagnano G, Russo S, Saggini R, Vulpiani MC, Buselli P: Extracorporeal shockwaves as regenerative therapy in orthopedic traumatology: a narrative review from basic research to clinical practice. J Biol Regul Homeost Agents 2016;30:323–332.

23 Seil R, Wilmes P, Nührenbörger C: Extracorporeal shock wave therapy for tendinopathies. Expert Rev Med Devices 2006;3:463–470.

24 Wang CJ, Chen HS, Chen WS, Chen LM: Treatment of painful heels using extracorporeal shock wave. J Formos Med Assoc 2000;99:580–583.

25 Ko JY, Chen HS, Chen LM: Treatment of lateral epicondylitis of the elbow with shock waves. Clin Orthop Relat Res 2001;387:60–67.

26 Ogden JA, Tóth-Kischkat A, Schultheiss R: Principles of shock wave therapy. Clin Orthop Relat Res 2001;387:8–17.

27 Ogden JA, Alvarez RG, Marlow M: Shockwave therapy for chronic proximal plantar fasciitis: a meta-analysis. Foot Ankle Int 2002;23:301–308.

28 Wang CJ: An overview of shock wave therapy in musculoskeletal disorders. Chang Gung Med J 2003;26:220–232.

29 Ogden JA, Alvarez RG, Levitt RL, Johnson JE, Marlow ME: Electrohydraulic high-energy shock-wave treatment for chronic plantar fasciitis. J Bone Joint Surg Am 2004;86:2216–2228.

30 Furia JP: Safety and efficacy of extracorporeal shock wave therapy for chronic lateral epicondylitis. Am J Orthop (Belle Mead NJ) 2005;34:13–19; discussion 19.

31 Radwan YA, ElSobhi G, Badawy WS, Reda A, Khalid S: Resistant tennis elbow: shock-wave therapy versus percutaneous tenotomy. Int Orthop 2008;32:671–677.

32 Rebuzzi E, Coletti N, Schiavetti S, Giusto F: Arthroscopy surgery versus shock wave therapy for chronic calcifying tendinitis of the shoulder. J Orthop Traumatol 2008;9:179–185.

33 Notarnicola A, Moretti L, Tafuri S, Forcignanò M, Pesce V, Moretti B: Reduced local perfusion after shock wave treatment of rotator cuff tendinopathy. Ultrasound Med Biol 2011;37:417–425.

34 Wang CJ: Extracorporeal shockwave therapy in musculoskeletal disorders. J Orthop Surg Res 2012;7:11.

35 Notarnicola A, Moretti B: The biological effects of extracorporeal shock wave therapy (eswt) on tendon tissue. Muscles Ligaments Tendons J 2012;2:33–37.

36 Notarnicola A, Pesce V, Vicenti G, Tafuri S, Forcignanò M, Moretti B: SWAAT study: extracorporeal shock wave therapy and arginine supplementation and other nutraceuticals for insertional Achilles tendinopathy. Adv Ther 2012;29:799–814.

37 Lee SS, Kang S, Park NK, Lee CW, Song HS, Sohn MK, Cho KH, Kim JH: Effectiveness of initial extracorporeal shock wave therapy on the newly diagnosed lateral or medial epicondylitis. Ann Rehabil Med 2012;36:681–687.

38 Aqil A, Siddiqui MR, Solan M, Redfern DJ, Gulati V, Cobb JP: Extracorporeal shock wave therapy is effective in treating chronic plantar fasciitis: a meta-analysis of RCTs. Clin Orthop Relat Res 2013;471:3645–3652.

39 Al-Abbad H, Simon JV: The effectiveness of extracorporeal shock wave therapy on chronic Achilles tendinopathy: a systematic review. Foot Ankle Int 2013;3:33–41.

40 Romeo P, Lavanga V, Pagani D, Sansone V: Extracorporeal shock wave therapy in musculoskeletal disorders: a review. Med Princ Pract 2014;23:7–13.

41 Saggini R, Di Stefano A, Saggini A, Bellomo RG: Clinical application of shock wave therapy in musculoskeletal disorders: part II related to myofascial and nerve apparatus. J Biol Regul Homeost Agents 2015; 29:771–785.

42 Gerdesmeyer L, Mittermayr R, Fuerst M, Al Muderis M, Thiele R, Saxena A, Gollwitzer H: Current evidence of extracorporeal shock wave therapy in chronic Achilles tendinopathy. Int J Surg 2015;24(pt B):154–159.

43 Gollwitzer H, Saxena A, DiDomenico LA, Galli L, Bouché RT, Caminear DS, Fullem B, Vester JC, Horn C, Banke IJ, Burgkart R, Gerdesmeyer L: Clinically relevant effectiveness of focused extracorporeal shock wave therapy in the treatment of chronic plantar fasciitis: a randomized, controlled multicenter study. J Bone Joint Surg Am 2015;97:701–708.

44 Leal C, Ramon S, Furia J, Fernandez A, Romero L, Hernandez-Sierra L: Current concepts of shockwave therapy in chronic patellar tendinopathy. Int J Surg 2015;24(pt B):160–164.

45 Moya D, Ramón S, Guiloff L, Gerdesmeyer L: Current knowledge on evidence-based shockwave treatments for shoulder pathology. Int J Surg 2015;24(pt B):171–178.

46 Thiele S, Thiele R, Gerdesmeyer L: Lateral epicondylitis: this is still a main indication for extracorporeal shockwave therapy. Int J Surg 2015;24(pt B):165–170.

47 Moya D, Ramón S, d'Agostino MC, Leal C, Aranzabal JR, Eid J, Schaden W: Incorrect methodology may favor ultrasound-guided needling over shock wave treatment in calcific tendinopathy of the shoulder. J Shoulder Elbow Surg 2016;25:e241–e243.

48 Arirachakaran A, Boonard M, Yamaphai S, Prommahachai A, Kesprayura S, Kongtharvonskul J: Extracorporeal shock wave therapy, ultrasound-guided percutaneous lavage, corticosteroid injection and combined treatment for the treatment of rotator cuff calcific tendinopathy: a network meta-analysis of RCTs. Eur J Orthop Surg Traumatol 2017;27:381–390.

49 Saggini R, Figus A, Troccola A, Cocco V, Saggini A, Scuderi N: Extracorporeal shock wave therapy for management of chronic ulcers in the lower extremities. Ultrasound Med Biol 2008;34:1261–1271.

50 Qureshi AA, Ross KM, Ogawa R, Orgill DP: Shock wave therapy in wound healing. Plast Reconstr Surg 2011;128:721e–727e.

51 Mittermayr R, Antonic V, Hartinger J, Kaufmann H, Redl H, Téot L, Stojadinovic A, Schaden W: Extracorporeal shock wave therapy (ESWT) for wound healing: technology, mechanisms, and clinical efficacy. Wound Repair Regen 2012;20:456–465.

52 Fioramonti P, Cigna E, Onesti MG, Fino P, Fallico N, Scuderi N: Extracorporeal shock wave therapy for the management of burn scars. Dermatol Surg 2012; 38:778–782.

53 Vardi Y, Appel B, Jacob G, Massarwi O, Gruenwald I: Can low-intensity extracorporeal shockwave therapy improve erectile function? A 6-month follow-up Pilot study in patients with organic erectile dysfunction. Eur Urol 2010;58:243–248.

54 Gruenwald I, Appel B, Kitrey ND, Vardi Y: Shockwave treatment of erectile dysfunction. Ther Adv Urol 2013;5:95–99.

55 Lei H, Liu J, Li H, Wang L, Xu Y, Tian W, Lin G, Xin Z: Low-intensity shock wave therapy and its application to erectile dysfunction. World J Mens Health 2013;31:208–214.

56 Abu-Ghanem Y, Kitrey ND, Gruenwald I, Appel B, Vardi Y: Penile low-intensity shock wave therapy: a promising novel modality for erectile dysfunction. Korean J Urol 2014;55:295–299.

57 Olsen AB, Persiani M, Boie S, Hanna M, Lund L: Can low-intensity extracorporeal shockwave therapy improve erectile dysfunction? A prospective, randomized, double-blind, placebo-controlled study. Scand J Urol 2014;49:329–333.

58 Bechara A, Casabé A, De Bonis W, Nazar J: Effectiveness of low-intensity extracorporeal shock wave therapy on patients with Erectile Dysfunction (ED) who have failed to respond to PDE5i therapy. A Pilot study. Arch Esp Urol 2015;68:152–160.

59 Chung E, Cartmill R: Evaluation of clinical efficacy, safety and patient satisfaction rate after low-intensity extracorporeal shockwave therapy for the treatment of male erectile dysfunction: an Australian first open-label single-arm prospective clinical trial. BJU Int 2015;11(suppl 5):46–49.

60 Shimokawa H, Ito K, Fukumoto Y, Yasuda S: Extracorporeal cardiac shock wave therapy for ischemic heart disease. Shock Waves 2008;17:449–455.

61 Wang Y, Guo T, Cai HY, Ma TK, Tao SM, Sun S, Chen MQ, et al: Cardiac shock wave therapy reduces angina and improves myocardial function in patients with refractory coronary artery disease. Clin Cardiol 2010;33:693–699.

62 Ito K, Fukumoto Y, Shimokawa H: Extracorporeal shock wave therapy for ischemic cardiovascular disorders. Am J Cardiovasc Drugs 2011;11:295–302.

63 Yang P, Peng Y-Z, Guo T, Wang Y, Cai HY, Zhou P: Clinical study of the extracorporeal cardiac shock wave therapy for coronary artery disease. Heart BMJ 2012;98:s2.

Vincent · d'Agostino

64 Kaller M, Faber L, Bogunovic N, Horstkotte D, Burchert W, Lindner O: Cardiac shock wave therapy and myocardial perfusion in severe coronary artery disease. Clin Res Cardiol 2015;104:843–849.

65 Holfeld J, Lobenwein D, Tepekoylu C, Grimm M: Shockwave therapy of the heart. Int J Surg 2015;24: 165–170.

66 Vainer J, Habets JH, Schalla S, Lousberg AH, de Pont CD, Voo SA, et al: Cardiac shockwave therapy in patients with chronic refractory angina pectoris. Neth Heart J 2016;24:343–349.

67 Yamaya S, Ozawa H, Kanno H, Kishimoto KN, Sekiguchi A, Tateda S, et al: Low-energy extracorporeal shock wave therapy promotes vascular endothelial growth factor expression and improves locomotor recovery after spinal cord injury. J Neurosurg 2014; 121:1514–1525.

68 Lee JH, Kim SG: Effects of extracorporeal shock wave therapy on functional recovery and neurotrophin-3 expression in the spinal cord after crushed sciatic nerve injury in rats. Ultrasound Med Biol 2015;41:790–796.

69 Lobenwein D, Tepekoylu C, Kozaryn R, Pechriggl EJ, Bitsche M, Graber M, et al: Shock wave treatment protects from neuronal degeneration via a toll-like receptor 3 dependent mechanism: implications of a first-ever causal treatment for ischemic spinal cord injury. J Am Heart Assoc 2015;4:e002440.

70 Wang L, Jiang Y, Jiang Z, Han L: Effect of low-energy extracorporeal shock wave on vascular regeneration after spinal cord injury and the recovery of motor function. Neuropsychiatr Dis Treat 2016;12:2189–2198.

71 Vincent KC, Schaden W, Karalus LA, Craig JH, Poratt D: Acoustic stimulation and tropism on skeletal muscles: tissue resilience and regeneration in sports and ageing. 19th Scientific Congress of the International Society for Medical Shockwave Treatment, 13th–15th July, 2016, Kuching, Malaysia.

72 Vincent KC: ESWT Influenced Musculogenesis in an Athletic and Non-Athletic Population. 13th National Congress of the S.I.T.O.D, 13th–14th, Naples, Italy, October, 2016.

73 Rassweiler JJ, Knoll T, Kohrmann KU: Shock wave technology and application: an update. Eur Urol 2011;59:784–796.

74 International Society for Medical Shockwave Treatment (2014). ISMST Consensus Statement: Terms and Definitions [online], available: https://www.shockwavetherapy.org/fileadmin/user_upload/dokumente/PDFs/Formulare/ismst-consensus-statement-terms-and-definition-2008.pdf [accessed 08 March, 2017].

75 Schmitz C, Csaszar NB, Milz S, Schieker M, Maffulli N, Rompe JD, Furia JP: Efficacy and safety of extracorporeal shock wave therapy for orthopedic conditions: a systematic review on studies listed in the PEDro database. Br Med Bull 2015;116:115–138.

76 International Society for Medical Shockwave Treatment. ISMST Recommendations: Physics recommendations-Basic physical principals [online]. https://www.shockwavetherapy.org/about-eswt/ismst-recommendations/ (accessed March 17, 2017).

77 d'Agostino MC, Craig K, Tibalt E, Respizzi S: Shock wave as biological therapeutic tool: from mechanical stimulation to recovery and healing, through mechanotransduction. Int J Surg 2015;24(pt B):147–153.

78 Gerdesmeyer L, Maier M, Haake M, Schmitz C: Physical-technical principles of extracorporeal shockwave therapy (ESWT). Orthopade 2002;31: 610–617.

79 Gerdesmeyer L, Henne M, Gobel M, Diehl P: Physical principles and generation of Shock Waves; in Gerdesmeyer L, Weil LS (eds): Extracorporeal Shock Wave Therapy: Clinical Results, Technologies, Basics. Towson, Data Trace Publishing Company, 2006, pp 11–20.

80 Speed C: A systematic review of shockwave therapies in soft tissue conditions: focusing on the evidence. Br J Sports Med 2014;48:1538–1542.

81 Novak P: Physics: F-SW and R-SW. Basic information on focused and radial shock wave physics; in Lohrer H, Gerdesmeyer L (eds): Multidisciplinary Medical Applications. Heilbronn, Daniela Bamberg, 2015, pp 28–49.

82 Lohrer H, Nauck T, Korakakis V, Malliaropoulos N: Historical ESWT paradigms are overcome: a narrative review. Biomed Res Int 2016;2016:3850461

83 Ramon S, Gleitz M, Hernandez L, Romero LD: Update on the efficacy of extracorporeal shockwave treatment for myofascial pain syndrome and fibromyalgia. Int J Surg 2015;24(pt B):201–206.

84 Cleveland RO, Chitnis PV, McClure SR: Acoustic field of a ballistic shock wave therapy device. Ultrasound Med Biol 2007;33:1327–1335.

85 van der Worp H, van den Akker-Scheek I, van Schie H, Zwerver J: ESWT for tendinopathy: technology and clinical implications. Knee Surg Sports Traumatol Arthrosc 2013;21:1451–1458.

86 Haake M, Konig IR, Decker T, Riedel C, Buch M, Muller HH; Extracorporeal Shock Wave Therapy Clinical Trial Group: Extracorporeal shock wave therapy in the treatment of lateral epicondylitis: a randomized multicenter trial. J Bone Joint Surg Am 2002;84:1982–1991.

87 Buchbinder R, Ptasznik R, Gordon J, Buchanan J, Prabaharan V, Forbes A: Ultrasound-guided extracorporeal shock wave therapy for plantar fasciitis: a randomized controlled trial. JAMA 2002;288:1364–1372.

88 Tepekoylu C, Wang FS, Kozaryn R, Albrecht-Schgoer K, Theurl M, Schaden W, et al: Shock wave treatment induces angiogenesis and mobilizes endogenous CD31/CD34-positive endothelial cells in a hindlimb ischemia model: implications for angiogenesis and vasculogenesis. J Thorac Cardiovas Surg 2013;146:971–978.

89 Sukubo NG, Tibalt E, Respizzi S, Locati M, d'Agostino MC: Effect of shock waves on macrophages: a possible role in tissue regeneration and remodeling. Int J Surg 2015;24:124–130.

90 Holfeld J, Tepekoylu C, Ressig C, Lobenwein D, Scheller B, Kirchmair E, el at: Toll-like receptor 3 signalling mediates angiogenic response upon shockwave treatment of ischaemic muscle. Cardiovasc Res 2016;109:331–343.

91 Notarnicola A, Moretti L, Tafuri S, Panella A, Filiponi M, Casalino A, et al: Shockwave therapy in the management of complex regional pain syndrome in medial femoral condyle of the knee. Ultrasound Med Biol 2010;36:874–879.

92 Vincent KC: Medical Shockwaves a Treatment Option for Complex and Neuropathic Pain Syndromes? A Compilation of Case Reports. 17th Scientific Congress of the International Society for Medical Shockwave Treatment, 26th–28th, Milan, Italy, 2014.

93 Richardson JD, Vasko MR: Cellular mechanisms of neurogenic inflammation. J Pharmacol Exp Ther 2002;302:839–845.

94 Vetrano M, d'Alessandro F, Torrisi MR, Ferretti A, Vulpiani MC, Visco V: Extracorporeal shock wave therapy promotes cell proliferation and collagen synthesis of primary cultured human tenocytes. Knee Surg Sports Traumatol Arthrosc 2011;19: 2159–2168.

95 Visco V, Vulpiani MC, Torrisi MR, Ferretti A, Pavan A, Vetrano M: Experimental studies on the biological effects of extracorporeal shock wave therapy on tendon models. A review of the literature. Muscles Ligaments Tendons J 2014;4:357–361.

96 Mariotto S, de Prati AC, Cavalieri E, Amelio E, Marlinghaus E, Suzuki H: Extracorporeal shock wave therapy in inflammatory diseases: molecular mechanism that triggers anti-inflammatory action. Curr Med Chem 2009;16:2366–2372.

97 Yahata K, Kanno H, Ozawa H, Yamaya S, Tateda S, Ito K, et al: Low-energy extracorporeal shock wave therapy for promotion of vascular endothelial growth factor expression and angiogenesis and improvement of locomotor and sensory functions after spinal cord injury. J Neurosurg Spine 2016;25:645–755.

98 Stojadinovic A, Elster EA, Anam K, Tadaki D, Amare M, Zins S, Davis TA: Angiogenic response to extracorporeal shock wave treatment in murine skin isografts. Angiogenesis 2008;11:369–380.

99 Sansone V, D'Agostino MC, Bonora C, Sizzano F, De Girolamo L, Romeo P: Early angiogenic response to shock waves in a three-dimensional model of human microvascular endothelial cell culture (HMEC-1). J Biol Regul Homeost Agents 2012;26:29–37.

100 Tamma R, dell'Endice S, Notarnicola A, Moretti L, Patella S, Patella V, Zallone A, Moretti B: Extracorporeal shock waves stimulate osteoblast activities. Ultrasound Med Biol 2009;35:2093–2100.

101 Hausdorf J, Sievers B, Schmitt-Sody M, Jansson V, Maier M, Mayer-Wagner S: Stimulation of bone growth factor synthesis in human osteoblasts and fibroblasts after extracorporeal shock wave application. Arch Orthop Trauma Surg 2011;131:303–309.

102 Kearney CJ, Hsu HP, Spector M: The use of extracorporeal shock wave-stimulated periosteal cells for orthotopic bone generation. Tissue Eng Part A 2012;18:1500–1508.

103 Suhr F, Delhasse Y, Bungartz G, Schmidt A, Pfannkuche K, Bloch W: Cell biological effects of mechanical stimulations generated by focused extracorporeal shock wave applications on cultured human bone marrow stromal cells. Stem Cell Res 2013;11:951–964.

104 ShI X, Garry DJ: Muscle stem cells in development, regeneration, and disease. Genes Dev 2006;20: 1692–1708.

105 Christov C, Chretien F, Abou-Khalil R, Bassez G, Vallet G, Authier FJ, et al: Muscle satellite cells and endothelial cells: close neighbors and privileged partners. Mol Biol Cell 2007;18:1397–1409.

106 Mounier R, Chretien F, Chazaud B: Blood vessels and the satellite cell niche. Curr Top Dev Biol 2011; 96:121–138.

107 Ceafalan LC, Popescu BO, Hinescu ME: Cellular players in skeletal muscle regeneration. Bio Med Res Int 2014;2014:957014.

108 Elsafadi M, Manikandan M, Dawud RA, Alajez NM, Hamam R, Alfayez M, et al: Transgelin is a TGFβ-inducible gene that regulates osteoblastic and adipogenic differentiation of human skeletal stem cells through actin cytoskeleston organization. Cell Death Dis 2016;7:e2321.

109 Bentzinger CF, Wang YX, Dumont NA, Rudnicki MA: Cellular dynamics in the muscle satellite cell niche. EMBO Rep 2013;14:1062–1072.

110 Pevsner-Fischer M, Morad V, Cohen-Sfady M, Russso-Noori L, Zanin-Zhorov A, Cohen IR, et al: Toll-like receptors and their ligands control mesenchymal stem cell functions. Blood 2007;109:1422–1432.

111 Qi C, Xiaofeng X, Xiaoguang W: Effects of Toll-Like Receptors 3 and 4 in the Osteogenesis of Stem Cells. Stem Cell Int 2014;2014:917168.

112 Delarosa O, Dalemans W, Lombardo E: Toll-like receptors as modulators of mesenchymal stem cells. Front Immunol 2012;3:182.

113 Shirjang S, Mansoori B, Solali S, Hagh MF, Shamsasenjan K: Toll-like receptors as a key regulator of mesenchymal stem cell function: an up-to-date review. Cell Immunol 2017;315:1–10.

114 Alvarado AG, Lathia JD: Taking a toll on self-renewal: TLR-mediated innate immune signaling in stem cells. Trends Neurosci 2016;39:463–471.

115 Ingber DE: Mechanobiology and diseases of mechanotransduction. Ann Med 2003;35:564–577.

116 Orr AW, Helmke BP, Blackman BR, Schwartz MA: Mechanisms of Mechanotransduction. Dev Cell 2006;10:11–20.

117 Huang H, Kamm RD, Lee RT: Cell mechanics and mechanotransduction: pathways, probes, and physiology. Am J Physiol 2004;287:C1–C11.

118 Chen CS: Mechanotransduction – a field pulling together? J Cell Sci 2008;121:3285–3292.

119 Chiquet M, Gelman L, Lutz R, Maier S: From mechanotransduction to extracellular matrix gene expression in fibroblasts. Biochim Biophys Acta 2009; 1793:911–920.

120 Jaalouk DE, Lammerding J: Mechantransduction gone awry. Nat Rev Mol Cell Biol 2009;10:63–73.

121 Paluch EK, Nelson CM, Biais N, Fabry B, Moller J, Pruitt BL, et al: Mechanotransduction: use the force(s). BMC Biol 2015;4:47.

122 Chopra A, Murray ME, Byfield FJ, Mendez MG, Halleluyan R, Restle DJ, et al: Augmentation of integrin-mediated mechanotransduction by hyaluronic acid. Biomaterials 2014;35:71–82.

123 Schwartz MA: Integrins and extracellular matrix in mechanotransduction. Cold Spring Harb Perspect Biol 2010;2:e005066.

124 Sun Z, Guo SS, Fassler R: Integrin-mediated mechanotransduction. J Cell Biol 2016;215:445–456.

125 du Roure O, Saez A, Buguin A, Austin RH, Chavrier P, Silberzan P, Ladoux B: Force mapping in epithelial cell migration. Proc Natl Acad Sci USA 2005; 102:2390–2395.

126 Engler AJ, Sen S, Sweeney HL, Discher DE: Matrix elasticity directs stem cell lineage specification. Cell 2006;126:677–689.

127 Vogel V, Sheetz MP: Cell fate regulation by coupling mechanical cycles to biochemical signaling pathways. Curr Opin Cell Biol 2009;21:38–46.

128 Weaver VM, Petersen OW, Wang F, Larabell CA, Briand P, Damsky C, Bissell MJ: Reversion of the malignant phenotype of human breast cells in three-dimensional culture and in vivo by integrin blocking antibodies. J Cell Biol 1997;137:231–245.

129 You L, Cowin SC, Schaffler MB, Weinbaum S: A model for strain amplification in the actin cytoskeleton of osteocytes due to fluid drag on pericellular matrix. J Biomech 2001;34:1375–1386.

130 Alenghat FJ, Nauli SM, Kolb R, Zhou J, Ingber DE: Global cytoskeletal control of mechanotransduction in kidney epithelial cells. Exp Cell Res 2004; 301:23–30.

131 Han Y, Cowin SC, Schaffler MB, Weinbaum S: Mechanotransduction and strain amplification in osteocyte cell processes. Proc Natl Acad Sci U S A 2004;101:16689–16694.

132 Hausner T, Nogradi A: The Use of Shock Waves in Peripheral Nerve Regeneration: New Perspectives?; in Guena S, Perroteau I, Tos P, Battiston B (eds): International Review of Neurobiology: Tissue Engineering of The Peripheral Nerve: Biomaterials and Physical Therapy (vol 109). London, Academic Press2013, p 88.

133 Raabe O, Shell K, Goessl A, Crispens C, Delhasse Y, Eva A: Effect of extracorporeal shock wave on proliferation and differentiation of equine adipose tissue-derived mesenchymal stem cells in vitro. Am J Stem Cells 2013;2:62–773.

134 Leone L, Raffa S, Vetrano M, Ranieri D, Malisan F, Scrofani C, et al: Extracorporeal shock wave treatment (ESWT) enhances the in vitro-induced differentiation of human tendon-derived stem/progenitor cells (hTSPCs).Oncotarget 2016;7: 6410–6423.

135 Trompetto C, Avanzino L, Bove M, Marinelli L, Molfetta L, Trentini R, et al: External shock waves therapy in dystonia: preliminary results. Eur J Neurol 2009;16:517–521.

136 Mori L, Marinelli L, Pelosin E, Curra A, Molfetta L, Abbruzzese G, Trompetto C: Shock waves in the treatment of muscle hypertonia and dystonia. Biomed Res Int 2014;2014:637–450.

137 Lee JY, Kim SN, Lee IS, Jung H, Lee KS, Koh SE: Effects of extracorporeal shock wave therapy on spasticity in patients after brain injury: a meta-analysis. J Phy Ther Sci 2014;26:1641–1647.

138 Hala A, Gawad A, Karim AEA, Mohammad H: Shock wave therapy for spastic plantar flexor muscles in hemiplegic cerebral palsy children. Egyptian J Human Gen 2015;16:269–275.

139 Mirea A, Onose G, Padure L, Rosulescu E: Extracorporeal Shockwave Therapy (ESWT) benefits in spastic children with Cerebral Palsy (CP). J Med Life 2014;7:127–132.

140 Park DS, Kwon DR, Park GY, Lee MY: Therapeutic effect of extracorporeal shock wave therapy according to treatment session on gastrocnemius muscle spasticity in children with spastic cerebral palsy: a pilot study. Ann Rehabil Med 2015;39:914–921.

141 Vincent KC, d'Agostino MC, Poratt D, Walker M: Utilisation of eswt to restore peripheral vibrosensory perception in a non-sensate Type 1 diabetic foot. 15th Scientific Congress of the International Society for Medical Shockwave Treatment, 4th–7th, Cartegena, Columbia, 2012.

142 Chen YL, Chen KH, Yin TC, Huang TH, Yuen CM, Chung SY, et al: Extracorporeal shock wave therapy effectively prevented diabetic neuropathy. Am J Trans Res 2015;7:2543–2560.

143 Chen TW, Lin CW, Lee CL, Chen CH, Chen YJ, Lin TY, Huang MH: The efficacy of shock wave therapy in patients with knee osteoarthritis and popliteal cyamellla. Kaohsiung J Med Sci 2014;30:362–370.

144 Zhao Z, Jing R, Shi Z, Zhao B, Ai Q, Xing G: Efficacy of extracorporeal shockwave therapy for knee osteoarthritis: a randomized controlled trial. J Surg Res 2013;185:661–666.

145 Kim JH, Kim JY, Choi CM, Lee JK, Kee HS, Jung KI, Yoon SR: The dose-related effects of extracorporeal shock wave therapy for knee osteoarthritis. Ann Rehabil Med 2015;39:616–623.

146 Vincent KC, Takai B, Craig J, McDonald D: Medical shockwaves for chronic low back pain: a case series. 18th Scientific Congress of the International Society for Medical Shockwave Treatment, 16th–18th, Mendoza, Argentina, July 2015.

147 Freitag K: 2016, Uterine Fibroids and Shockwaves. 19th 18th Scientific Congress of the International Society for Medical Shockwave Treatment, 13th–15th July, 2017.

148 Craig K, d'Agostino C, Poratt D, Walker M: Original hypothesis: extracorporeal shockwaves as a homeostatic autoimmune restorative treatment (HART) for Type 1 diabetes mellitus. Med Hypotheses 2014;83:250–253.

Kenneth Craig S. Vincent, MD
Kompass Health Associates
27A Haughey Avenue
Three Kings 1042, Aukland (New Zealand)
E-Mail ken@kompasshealth.org

Vincent · d'Agostino

Wang C-J, Schaden W, Ko J-Y (eds): Shockwave Medicine.
Transl Res Biomed. Basel, Karger, 2018, vol 6, pp 17–26 (DOI: 10.1159/000485056)

Development of Extracorporeal Shockwave Therapy for Treatment of Osteonecrosis of the Femoral Head

Jai-Hong Cheng[a, c] · Shan-Ling Hsu[a, b] · Ching-Jen Wang[a, b]

[a]Center for Shockwave Medicine and Tissue Engineering, [b]Department of Orthopedic Surgery, and [c]Medical Research, Kaohsiung Chang Gung Memorial Hospital and Chang Gung University College of Medicine, Kaohsiung, Taiwan

Abstract

Osteonecrosis of the femoral head (ONFH), also known as avascular necrosis of the femoral head, is usually caused by the death of bone tissue due to disruption of blood supply. Multiple factors can induce ONFH such as drug abuse alcohol overuse overweight, trauma, corticosteroid, chemotherapy, and idiopathic. Symptomatic hips with early ONFH are treated conservatively, while joint replacement is indicated in late stage. For early stages, core decompression with or without bone grafting is recommended to preserve the femoral head. However, the results are inconsistent. Some achieved limited success but none universal. Some reported poor results that many patients eventually undergo hip replacement surgery. Total hip replacement in young patients is at risk for loosening and dislocation. Other methods of treatment included non-weight-bearing therapy, hyperbaric oxygen therapy, and extracorporeal shockwave therapy (ESWT). ESWT has been used to treat various tendinopathies and fractures with poor bone healing with reasonable success for 30–40 years. Recently, ESWT was investigated in the treatment of ONFH. ESWT is a noninvasive device that exerts beneficial effects of improving bone regeneration, anti-inflammatory, pain relief, and angiogenesis in the treatment of ONFH. Many experiments showed that ESWT is more effective than core decompression with or without bone grafting for patients with early stage of ONFH. The purpose of the present article is to review the clinical trials and animal studies of ESWT on ONFH with perspective and challenge in the future.

Osteonecrosis of the Femoral Head and Extracorporeal Shockwave Therapy

The etiology of osteonecrosis of the femoral head (ONFH) is multifactorial including alcohol abuse, overweight, trauma, corticosteroid usage, chemotherapy, thalassemia, pregnancy, HIV infection, organ transplantation, ischemia, systemic lupus erythematosus (SLE), and idiopathic [1, 2]. ONFH is usually progressive that leads to collapse or fracture of femoral head, and the disease finally leads to the femoral head collapse, destruction and degenerative changes of hip joint. Several prognostic factors could be observed for the progression of ONFH including the extent and location of the osteonecrotic lesion, lesion size, bone marrow edema in the proximal femur, and genetic factors [2, 3]. Imaging studies, particularly MRI, are the major methods that can accurately evaluate and assess the progression of ONFH. These imaging studies may also assess the risks of femoral head collapse from patients' perspective and clarify the patient expectation from the perspective of the history of the natural development of the disease. There are several classification systems to identify and staging the ONFH, including Ficat and Arlet classification, Steinberg-the University of Pennsylvania staging system, Association Research Circulation Osseous (ARCO), and the Japanese Orthopaedic Association classification systems (Table 1) [4]. These classification systems provide information of ONFH on the prognosis and assist with treatment decision.

There are many options for treating ONFH including non-weight-bearing (NWB) therapy, pharmacologic treatment, physical support, hyperbaric oxygen (HBO) therapy, core decompression, extracorporeal shockwave therapy (ESWT), and surgery [5]. The purpose of treatment for ONFH is to prevent collapse or delay the stage progression of the femoral head. However, the optimal treatment for ONFH is still not well defined and remains controversial. In general, patients with early ONFH prefer conservative nonsurgical management, but unsatisfactory functional outcomes were noticed in most patients. There is still no completely effective method, even with regard to surgery for ONFH. Better results often rely on early diagnosis and treatment. The type of surgery is dependent on the stage of ONFH images [6, 7]. The operative treatment such as the core decompression with or without bone grafting is the most common surgical procedure for precollapse of ONFH and is regarded as the gold standard for preserving the femoral head. However, the potential risk included infections and immune response, and reports unsatisfactory results [8]. Recently, ESWT was reported effective by researchers in the treatment of ONFH [9–11]. This article describes a new effective and noninvasive method of ESWT for the treatment of ONFH.

ESWT for ONFH

Researchers reported that focused ESWT is effective in the treatment of ONFH [9–12]. However, the exact mechanism of ESWT in ONFH is still unknown. In 2001, Ludwig et al. [9] first reported ESWT treatment in 22 patients with ONFH of ACRO

Ficat and Arlet classification	Steinberg-the University of Pennsylvania staging system	Association Research Circulation Osseous	Japanese Orthopaedic Association classification systems
Stage description (Plain radiographs, MRI and clinical features)	Stage description (the type and extent of radiologic and pathologic changes)	Stage description (Radiographs, computed tomography (CT), bone scans, and MRI)	Stage description (Radiographs, bone scans, and MRI)
Stage 0: A. Plain radiograph: normal B. MRI: normal C. Clinical symptoms: nil	Stage 0: Normal or nondiagnostic radiograph, bone scan, and MRI	Stage 0: *Radiograph:* Normal *CT:* Normal *Bone scan:* Normal *MRI:* Normal *Histology:* Plasmostasis and marrow necrosis *Clinical:* Normally no pain	Satge1: There are no specific findings of osteonecrosis on x-ray images, although specific findings are observed on MRI, bone scintigram, or histology
Stage I: A. Plain radiograph: normal or minor osteopenia B. MRI: edema C. Bone scan: increased uptake D. Clinical symptoms: pain typically in the groin	Stage 1: Normal radiograph; abnormal bone scan and MRI A. Mild B. Moderate C. Severe	Stage 1: *Radiograph:* Normal *CT:* Normal *Bone scan:* Cold spot *MRI:* Necrotic area and reactive interface *Histology:* Bone necrosis and inflammatory response *Clinical:* Normally no pain	Stage 2: The demarcating sclerosis is observed without collapse of the femoral head
Stage II: A. Plain radiograph: mixed osteopenia and/or sclerosis and/or subchondral cysts, without any subchondral lucency (crescent sign: see below) B. MRI: geographic defect C. Bone scan: increased uptake D. Clinical symptoms: pain and stiffness	Stage 2: Lucent and sclerotic changes in femoral head A. Mild B. Moderate C. Severe	Stage 2: *Radiograph:* Mottled and sclerotic rim *CT:* Mottled and sclerotic rim *Bone scan:* Cold in hot spot *MRI:* Necrotic area and reactive interface *Histology:* Reactive interface repair *Clinical:* May have pain	Stage 3: The collapse of the femoral head, including crescent sign, is observed without joint-space narrowing; mild osteophyte formation in the femoral head or acetabulum may be observed Stage 3A. The collapse of the femoral head is less than 3 mm Stage 3B. The collapse of the femoral head is 3 mm or greater
Stage III: A. Plain radiograph: crescent sign and eventual cortical collapse B. MRI: same as plain film C. Clinical symptoms: pain and stiffness +/– radiation to knee and limp	Stage 3: Subchondral collapse (crescent sign) without flattening A. Mild B. Moderate C. Severe	Stage 3: *Radiograph:* Crescent sign and/or flattening *CT:* Subchondral fracture *Bone scan:* Cold in hot spot *MRI:* Crescent sign *Histology:* Resorption and subchondral fracture *Clinical:* Pain	Stage 4: The osteoarthritic changes are observed
Stage IV: A. Plain radiograph: end stage with evidence of secondary degenerative change B. MRI: same as plain radiograph C. Clinical symptoms: pain and limp	Stage 4: Flattening of femoral head A. Mild B. Moderate C. Severe	Stage 4: *Radiograph:* Collapse *CT:* Collapse *Bone scan:* Hot in hot spot *MRI:* Collapse *Histology:* Flattening and cartilage destruction *Clinical:* Pain	
	Stage 5: Joint narrowing, acetabular changes A. Mild (average of femoral head involvement) B. Moderate (as determined in stage 4 and estimated) C. Severe (acetabular involvement)		
	Stage 6: Advanced degenerative changes		

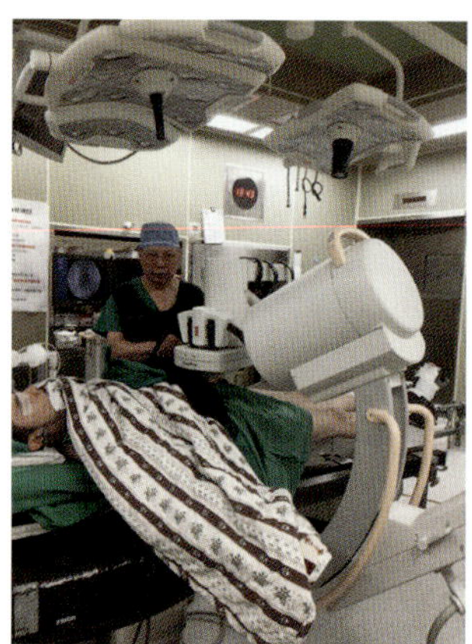

Fig. 1. A photograph showing shockwave treatment of ONFH by Professor Wang.

classification from stages 1–3 and were followed up for 1 year. The patients showed significant improvement in the visual analogue pain score, which decreased from 8.5 to 1.2. Harris hip score increased from 43.3 to 92 and the lesion size decreased or healed in 10 of 14 successfully treated patients. The therapeutic success rate was noticed in 14 patients, and 4 patients showed complete healing on MRI. Wang et al. [10] were the first to report that ESWT treated patients with ONFH by using FDA-approved machine OssaTron in Taiwan in 2005 (Fig. 1). They compared ESWT treatment in 23 patients (29 hips) with 25 patients (29 hips) receiving core decompression with nonvascularized fibular grafting with early stages of ONFH. The results showed significant improvement in pain and Harris hip scores ($p < 0.001$). There was a trend of reduction in the lesion size of ESWT group as compared to core decompression with fibular graft side on average follow-up of 25 months. In 2009, Chen et al. [13] compared the functional outcomes of ESWT on one side and total hip arthroplasty (THA) on the other hip in 17 patients with bilateral hip necrosis. The pain score and Harris hip score showed significant improvements after treatment ($p < 0.001$). A significant difference in the levels of improvement was observed between the 2 sides, with the difference favoring the ESWT side ($p < 0.001$). The results showed that 13 patients rated ESWT better than THA and only 4 patients reported comparable results between THA and ESWT and no one scored that THA was better than ESWT. At 1-year follow-up, of 36 patients with 42 hips with ONFH, the results showed 6 cases cured, 13 cases markedly improved, 16 cases improved, and 7 cases of failures [14].

The long-term follow-up of ESWT for ONFH was reported by Dr. Vulpiani et al. [11] in 2012 . The ESWT for treatment of 36 patients with ONFH was followed up at 3, 6, 12, and 24 months. The results were significantly associated with ARCO staging of the lesions after ESWT. Patients with early stage ONFH with ARCO stage I (100%) and stage II (81.8%) achieved excellent or good results than those in their late phase with ONFH with ARCO stage IIIa (26.7%) at follow-up ($p < 0.005$). Kusz demonstrated that ESWT resulted in considerable enhancement of quality of life in ONFH patients. Patients experienced pain reduction (visual analogue pain score decreased from 6.75 to 2.5) and increased mobility of the treated hip joint. Harris hip score increased from 55.21 to 89.21. However, they only followed up patients for 6 weeks [15]. In the same year, Wang et al. [16] reported long-term results of ESWT and core decompression in ONFH with 8- to 9-year follow-up. There were 48 patients with 57 hips in the study. The ESWT group consisted of 23 patients with 29 hips and the core decompression group had 25 patients with 28 hips. The functional results showed that 76% of hips were good or fair and 24% were poor after ESWT. On the other hand, 21% of hips were rated good or fair and 79% poor after core decompression. The results demonstrated that ESWT had better results than surgery in the treatment of early ONFH in long-term follow-up. Similar results were reported by Lee et al. [17]. They evaluated 24 patients with ARCO-staged ONFH in 32 hip joints that were treated with ESWT and follow-up from 1993 to 2012. The visual analogue scale scoring in group 1 (ARCO stages I and II) showed a median of 7–1.5 ($p < 0.001$) and group 2 (ARCO stage III) showed a mean of 7 to 4 ($p = 0.056$). In Harris hip score (HHS) analysis, group 1 showed significant improvement from 65.5 to 95 ($p < 0.001$), but the improvement was non-significant for group II ($p = 0.280$). The results indicated that ESWT was effective in early and mid-stage of ONFH. The largest patient population on this topic was reported by Gao et al. [18]. They showed a total of 335 patients with 528 hips treated with ESWT and followed up at 3, 6, and 12 months. The pain reduction ($p = 0.00006$) improved mobility of the treated hips ($p = 0.00091$), and bone marrow edema ($p = 0.007$) showed significant improvement after ESWT. The lesion size decreased after ESWT, but the differences were nonsignificant. Wang et al. [19] demonstrated that high dosage ESWT was more effective in the treatment of early-stage ONFH. They recruited 32 patients (42 hips) randomly and divided them into three groups. Group A (10 patients with 16 hips) received 2,000 impulses of ESWT at 0.510 mJ/mm^2 to each hip. Group B (11 patients with 14 hips) and group C (12 patients with 12 hips) received 4,000 and 6,000 impulses of ESWT to each hip respectively. The high-dosage group C showed significant improvement in pain score ($p = 0.037$) and Harris hip score ($p = 0.017$) than group A and B at 6 month follow-up. One case report showed radial ESWT improved joint effusion, bone density for treatment of ARCO stage IV ONFH [20]. Additional studies are needed to clarify the use of radial ESWT with particular reference to ONFH.

The Synergistic Effect and Cocktail Therapy of ESWT in ONFH Treatments

In 2003, 4 healthcare workers including 3 women and 1 man with an average age of 26 years were affected by severe acute respiratory syndrome and were treated with a massive dose of corticosteroids that resulted in the development of early ONFH [21]. Those patients were treated with cocktail therapy that consisted of ESWT, HBO therapy, and alendronate. The practical applications of treatments were processed on 4 patients with 8 hips. First, each hip was treated with 6,000 impulses of shockwave at

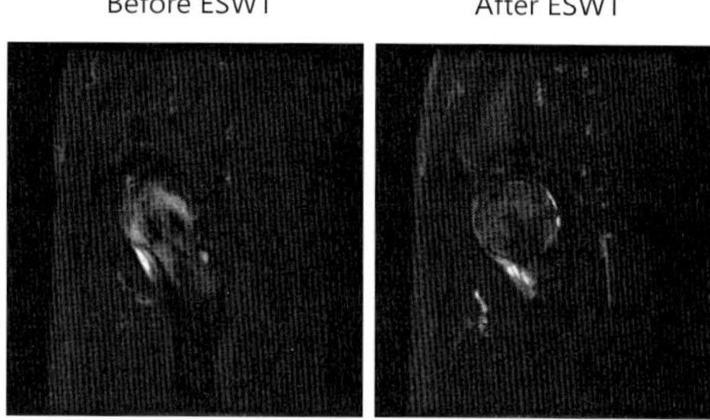

Fig. 2. The right hip before and after treatment showed the reduction of bone marrow edema and no further collapse; lesions of the femoral heads on MRI.

0.62 mJ/mm^2 energy flux density (EFD) in a single session. Then, HBO was performed once a day, 5 times a week for a total of 100 sessions. Patients received alendronate sodium 70 mg per week for 1 year. All patients returned to work as health care providers, and none of the hips required surgery during the 4 years follow-up period.

Our study also compared the effect of ESWT with and without alendronate for treatment of ONFH [22]. Hsu et al. [23] randomly divided forty-eight patients with 60 hips into group A (25 patients with 30 hips) and group B (23 patients with 30 hips). Group A was treated with 6,000 impulses of ESWT at 0.62 mJ/mm^2 to the affected hip in a single session. Group B with ESWT received alendronate 70 mg per week for 1 year. Both groups showed significant improvement in pain and function of the hip ($p < 0.001$), but the differences between the 2 groups were nonsignificant ($p = 0.400$ and $p = 0.313$ respectively). In 2010, a comparison of cocktail therapy and ESWT alone was made in 63 patients (98 hips). The results showed that there was no difference between combined ESWT alendronate and HBO and ESWT alone to treat early ONFH. Therefore, the synergistic effects of ESWT, HBO, and alendronate treatments were not observed in short-term follow-up.

ESWT Treatment of ONFH in SLE Patients

The standard methods for the treatment of patients with SLE usually included long-term corticosteroid. Patients with chronic corticosteroids therapy often caused the ONFH [24, 25]. In 2006, Lin et al. [26] showed a case report that ESWT was used effectively in the treatment of ONFH with SLE caused by corticosteroids therapy. The result showed pain score (score = 0) and Harris hip score (score = 100) of bilateral hips of ONFH, reduced bone marrow edema, and no progression. Collapse of the lesions was observed on MRI (Fig. 2). In a further study, 39 patients including 15 SLE (26 hips) and 24 non-SLE (29 hips) with ONFH were enrolled for ESWT (0.62 mJ/ mm^2; 6,000 impulses) treatment. The results showed that the therapeutic effects showed no difference in pain score ($p = 0.467$) and Harris hip score ($p = 0.194$) between SLE and non-SLE patients.

Dosage Does Matter and Challenges in Future

Presently, there are limited data describing the dose effect of shockwave on the biomechanical properties of ONFH. Previous studies demonstrated a dose-related effect of ESWT in musculoskeletal disorder [27–29]. Kong et al. [14] investigated the application value of extracorporeal shockwave shock to repair and reconstruct osseous tissue for the treatment of ONFH by using energy density of ESWT from 0.18 to 0.25 mJ/mm^2 in 36 patients with 42 hips. They found that ESWT had obvious therapeutic effects in the repairing and reconstructing osseous tissue and the hip Harris score and hip function had improved. Kusz et al. [15] assessed 9 patients with ONFH, ARCO stage I to III, treated by shockwave therapy with a dose of 1,500 pulses at an EFD of 0.4 mJ/mm^2 and a frequency of 4 Hz at each point, which each patient receiving 4 points in the femoral head and 5 therapy sessions. After 6 weeks follow-up, the patients demonstrated pain reduction and improved mobility of the treated joint with considerable improvement in quality of life. The effectiveness of ESWT was to reduce pain and slow down the progression of bone damage in 36 patients with unilateral ONFH of ACRO stage I to III. Each treatment included four sessions, with 2,400 impulses each administered at 0.50 mJ/mm^2 at 48–72 h intervals. Patients in ARCO stages I and II groups achieved significantly better results than those in the ARCO stage III group. ARCO stages I and II lesions were unchanged on radiographs and magnetic resonance images. Their results showed that ESWT in ARCO stages I and II may help to prevent progression of the area with avascular necrosis and manage pain [11].

The shockwave therapy with 6,000 impulses at 28 kV (0.62 mJ/mm^2) in a single session was used to treat 35 patients (47 hips) with ONFH [30]. Data showed that the local ESWT application resulted in significant systemic elevations of serum nitric oxide level, angiogenesis (vascular endothelial growth factor, VEGF; Von Willebrand factor, vWF; and fibroblast growth factor-basic) a decrease of transforming growth factor beta (TGF-β1), an increase of osteogenic markers (bone morphogenetic protein 2, BMP-2; osteocalcin, alkaline phosphatase and insulin-like growth factor), and a decrease of Dickkopf-related protein 1 (DKK1) and anti-inflammatory factors (soluble intercellular adhesion molecule and soluble vascular cell adhesion molecule, sVCAM) at 1 month after ESWT. In another study, they found the application of shockwave results in regeneration effects in 7 hips with ONFH with significantly more viable bone and less necrotic bone, higher cell concentration cell activity including phagocytosis and increases in vWF, VEGF, and CD 31 and Winless 3a and proliferating cell nuclear antigen, and decreases in VCAM and DKK1 [31]. When patients with SLE were affected by ONFH, ESWT also demonstrated its effectiveness with improvement in hip pain and function and image changes on X-ray and MRI in 15 SLE patients with 26 hips [32]. Besides the above, the effects of different dosages of ESWT were reported in early ONFH [19]. High-dose ESWT (6,000 impulses at 24 Kv [0.510 mJ/mm^2]) is more effective than low-dose energy (2,000 impulses at 24 Kv) in systemic beneficial effects such as enhancement of angiogenesis with improvement of microcirculation of the peri-necrotic areas. In turn, high-dose ESWT can improve subchondral bone remolding and prevent femoral head collapse more effectively than low-dose therapy in early ONFH. However, in 2016, Han et al. [33] reported lower energy density ESWT in the treatment of early stage of ONFH with 6-month follow-up. They showed ESWT indeed has a significant effect in treating the symptoms of ONFH, but there was no significant difference between the two groups (1,000 shocks/session, EFD per shock 0.12 and 0.32 mJ/mm^2, 4 sessions), regardless of the difference in EFD.

To summarize the results of previous researches, there is still no standard protocol of ESWT for clinical application in ONFH. Part of the reasons is due to the fact that the success rate of treatment in ONFH is more dependent on the stage of this disease. Many experimental and clinical researches tried to preserve the hip joint and prevent the progression to the secondary degenerative change during the early stage of ONFH. In order to set a standard procedure of shockwave therapy, the International Society for Medical Shockwave Treatment recommends a protocol, 0.62 mJ/mm^2 of EFD in 4,000 impulses delivered to the skin close to the damaged bone for ONFH treatment. Further studies are necessary to assess the optimal timing and dosage of ESWT on ONFH in the future.

In conclusion, ESWT is a new technology and has the potential of replacing surgery in ONFH without the surgical risks. In clinical trials and animal experiments, ESWT stimulates a cascade of signal transduction responses and a series of protein changes to exert its effectiveness in the management of ONFH through neovascularization and regeneration of the bone. Studies to date, however, have involved small numbers with short-term and long-term follow up. A protocol is needed to standardize the optimal timing and dosage of ESWT on ONFH. Further studies are necessary to assess this option in the future.

Disclosure Statement

The authors declared that they did not receive any honoraria or consultancy fees for writing this manuscript. No benefits in any form have been received or will be received from a commercial party related directly or indirectly to the subject of this article. One author (Ching-Jen Wang) serves as a member of the advisory committee of SANUWAVE (Suwanee, GA, USA), and this study was performed independent of the appointment. The remaining authors declared no financial conflicts of interest.

References

1 Bradway JK, Morrey BF: The natural history of the silent hip in bilateral atraumatic osteonecrosis. J Arthroplasty 1993;8:383–387.

2 Zalavras CG, Lieberman JR: Osteonecrosis of the femoral head: evaluation and treatment. J Am Acad Orthop Surg 2014;22:455–464.

3 Scaglione M, Fabbri L, Celli F, Casella F, Guido G: Hip replacement in femoral head osteonecrosis: current concepts. Clin Cases Miner Bone Metab 2015; 12(Suppl 1):51–54.

4 Cao H, Guan H, Lai Y, Qin L, Wang X: Review of various treatment options and potential therapies for osteonecrosis of the femoral head. J Orthop Translat 2016;4:57–70.

5 Wang C, Peng J, Lu S: Summary of the various treatments for osteonecrosis of the femoral head by mechanism: A review. Exp Ther Med 2014;8:700–706.

6 Hungerford DS: [Role of core decompression as treatment method for ischemic femur head necrosis]. Orthopade 1990;19:219–223.

7 Mont MA, Jones LC, Hungerford DS: Nontraumatic osteonecrosis of the femoral head: ten years later. J Bone Joint Surg Am 2006;88:1117–1132.

8 Mont MA, Carbone JJ, Fairbank AC: Core decompression versus nonoperative management for osteonecrosis of the hip. Clin Orthop Relat Res 1996;324:169–178.

9 Ludwig J, Lauber S, Lauber HJ, Dreisilker U, Raedel R, Hotzinger H: High-energy shock wave treatment of femoral head necrosis in adults. Clin Orthop Relat Res 2001;387:119–126.

10 Wang CJ, Wang FS, Huang CC, Yang KD, Weng LH, Huang HY: Treatment for osteonecrosis of the femoral head: comparison of extracorporeal shock waves with core decompression and bone-grafting. J Bone Joint Surg Am 2005;87:2380–2387.

11 Vulpiani MC, Vetrano M, Trischitta D, Scarcello L, Chizzi F, Argento G, Saraceni VM, Maffulli N, Ferretti A: Extracorporeal shock wave therapy in early osteonecrosis of the femoral head: prospective clinical study with long-term follow-up. Arch Orthop Trauma Surg 2012;132:499–508.

12 Zhang Q, Liu L, Sun W, Gao F, Cheng L, Li Z: Extracorporeal shockwave therapy in osteonecrosis of femoral head: a systematic review of now available clinical evidences. Medicine (Baltimore) 2017;96:e5897.

13 Chen JM, Hsu SL, Wong T, Chou WY, Wang CJ, Wang FS: Functional outcomes of bilateral hip necrosis: total hip arthroplasty versus extracorporeal shockwave. Arch Orthop Trauma Surg 2009;129:837–841.

14 Kong FR, Liang YJ, Qin SG, Li JJ, Li XL: [Clinical application of extracorporeal shock wave to repair and reconstruct osseous tissue framework in the treatment of avascular necrosis of the femoral head (ANFH)]. Zhongguo Gu Shang 2010;23:12–15.

15 Kusz D, Franek A, Wilk R, Dolibog P, Blaszczak E, Wojciechowski P, Krol P, Dolibog P, Kusz B: The effects of treatment the avascular necrosis of the femoral head with extracorporeal focused shockwave therapy. Ortop Traumatol Rehabil 2012;14:435–442.

16 Wang CJ, Huang CC, Wang JW, Wong T, Yang YJ: Long-term results of extracorporeal shockwave therapy and core decompression in osteonecrosis of the femoral head with eight- to nine-year follow-up. Biomed J 2012;35:481–485.

17 Lee JY, Kwon JW, Park JS, Han K, Shin WJ, Lee JG, Lee BH: Osteonecrosis of femoral head treated with extracorporeal shock wave therapy: analysis of short-term clinical outcomes of treatment with radiologic staging. Hip Pelvis 2015;27:250–257.

18 Gao F, Sun W, Li Z, Guo W, Wang W, Cheng L, Wang B: High-energy extracorporeal shock wave for early stage osteonecrosis of the femoral head: a single-center case series. Evid Based Complement Alternat Med 2015;2015:468090.

19 Wang CJ, Huang CC, Yip HK, Yang YJ: Dosage effects of extracorporeal shockwave therapy in early hip necrosis. Int J Surg 2016;35:179–186.

20 Ma YW, Jiang DL, Zhang D, Wang XB, Yu XT: Radial extracorporeal shock wave therapy in a person with advanced osteonecrosis of the femoral head. Am J Phys Med Rehabil 2016;95:e133–e139.

21 Wong T, Wang CJ, Hsu SL, Chou WY, Lin PC, Huang CC: Cocktail therapy for hip necrosis in SARS patients. Chang Gung Med J 2008;31:546–553.

22 Wang CJ, Wang FS, Yang KD, Huang CC, Lee MS, Chan YS, Wang JW, Ko JY: Treatment of osteonecrosis of the hip: comparison of extracorporeal shockwave with shockwave and alendronate. Arch Orthop Trauma Surg 2008;128:901–908.

23 Hsu SL, Wang CJ, Lee MS, Chan YS, Huang CC, Yang KD: Cocktail therapy for femoral head necrosis of the hip. Arch Orthop Trauma Surg 2010;130:23–29.

24 Dubois EL, Cozen L: Avascular (aseptic) bone necrosis associated with systemic lupus erythematosus. JAMA 1960;174:966–971.

25 Zizic TM, Marcoux C, Hungerford DS, Dansereau JV, Stevens MB: Corticosteroid therapy associated with ischemic necrosis of bone in systemic lupus erythematosus. Am J Med 1985;79:596–604.

26 Lin PC, Wang CJ, Yang KD, Wang FS, Ko JY, Huang CC: Extracorporeal shockwave treatment of osteonecrosis of the femoral head in systemic lupus erythematosis. J Arthroplasty 2006;21:911–915.

27 Wang CJ, Yang KD, Wang FS, Hsu CC, Chen HH: Shock wave treatment shows dose-dependent enhancement of bone mass and bone strength after fracture of the femur. Bone 2004;34:225–230.

28 Maier M, Tischer T, Milz S, Weiler C, Nerlich A, Pellengahr C, Schmitz C, Refior HJ: Dose-related effects of extracorporeal shock waves on rabbit quadriceps tendon integrity. Arch Orthop Trauma Surg 2002;122:436–441.

29 Kim JH, Kim JY, Choi CM, Lee JK, Kee HS, Jung KI, Yoon SR: The dose-related effects of extracorporeal shock wave therapy for knee osteoarthritis. Ann Rehabil Med 2015;39:616–623.

30 Wang CJ, Yang YJ, Huang CC: The effects of shockwave on systemic concentrations of nitric oxide level, angiogenesis and osteogenesis factors in hip necrosis. Rheumatol Int 2011;31:871–877.

31 Wang CJ, Wang FS, Ko JY, Huang HY, Chen CJ, Sun YC, Yang YJ: Extracorporeal shockwave therapy shows regeneration in hip necrosis. Rheumatology (Oxford) 2008;47:542–546.
32 Wang CJ, Ko JY, Chan YS, Lee MS, Chen JM, Wang FS, Yang KD, Huang CC: Extracorporeal shockwave for hip necrosis in systemic lupus erythematosus. Lupus 2009;18:1082–1086.
33 Han Y, Lee JK, Lee BY, Kee HS, Jung KI, Yoon SR: Effectiveness of Lower Energy Density Extracorporeal Shock Wave Therapy in the Early Stage of Avascular Necrosis of the Femoral Head. Ann Rehabil Med 2016;40:871–877.

Ching-Jen Wang, MD
Department of Orthopedic Surgery, Kaohsiung Chang Gung Memorial Hospital
123 Tai-Pei Road, Niao Sung District
Kaohsiung 83301 (Taiwan)
E-Mail w281211@adm.cgmh.org.tw

Cheng · Hsu · Wang

Wang C-J, Schaden W, Ko J-Y (eds): Shockwave Medicine.
Transl Res Biomed. Basel, Karger, 2018, vol 6, pp 27–41 (DOI: 10.1159/000485060)

Extracorporeal Shockwave Therapy for Tendinopathy

Jih-Yang Ko[a–d] · Feng-Sheng Wang[c, e, f]

[a]Center for Shockwave Medicine and Tissue Engineering, [b]Department of Orthopedic Surgery, and [c]Graduate Institute of Clinical Medical Sciences, Chang Gung University College of Medicine, Kaohsiung Chang Gung Memorial Hospital, Kaohsiung, Taiwan; [d]Department of Orthopaedic Surgery, Xiamen Chang Gung Hospital, Xiamen, China; [e]Department of Medical Research, and [f]Core Laboratory for Phenomics and Diagonistics, Kaohsiung Chang Gung Memorial Hospital, Kaohsiung, Taiwan

Abstract

Shockwaves are three-dimensional pressure pulses of microsecond duration with a peak pressure of 35–120 MPa and have been applied to the treatment of soft tissue and musculoskeletal disorders for over 20 years. The concentrated shockwave energy per unit area, the energy flux density (EFD, in mJ/mm^2), is currently used as a descriptive parameter of shockwave dosage. As a guideline, low energy extracorporeal SW therapy (ESWT) is EFD ≤0.12 mJ/mm^2, and high energy is >0.12 mJ/mm^2. Another commonly used grouping defines EFD <0.08 mJ/mm^2 as low energy, between 0.08 and 0.28 mJ/mm^2 as medium energy, and 0.28–0.6 mJ/mm^2 as high energy. ESWT is an efficacious and efficient alternative to surgery for rotator cuff calcification. Combining physical therapy, shockwaves treatment is effective and safe for chronic patellar tendinopathies. Recently, high-quality controlled trials have confirmed excellent results of ESWT in chronic Achilles tendinopathy. For proximal plantar fasciitis, the literature review unveiled some controversy in its effect, with the vast majority of papers favoring ESWT. The success rate ranged from 68 to 91% in shockwave treatment for lateral epicondylitis of the elbow. Further studies have to focus on the efficacy of different treatment modalities such as EFD, frequency, and the application pressure. © 2018 S. Karger AG, Basel

Introduction

Extracorporeal SW therapy (ESWT) has been applied to the treatment of soft tissue and musculoskeletal disorders for over 20 years. Shockwaves are three-dimensional pressure pulses of microsecond duration with a peak pressure of 35–120 MPa. The current form of medical ESWT involves focused shockwaves. These are concentrated into a small focal area of 2–8 mm diameter in order to optimize therapeutic effects while minimising effects on other tissues [1]. The mechanobiological effects are considered to depend on the energy delivered to a focal area. The concentrated shockwave energy per unit area, the energy flux density (EFD, in mJ/mm^2), is used to reflect the flow of shockwave energy in a perpendicular direction to the direction of propagation and currently used as descriptive parameters of shockwave dosage [2]. There remains no consensus as to the definition of "high-" and "low"-energy ESWT. However, as a guideline, low-energy ESWT is EFD ≤0.12 mJ/mm^2 and high-energy is >0.12 mJ/mm [1, 3]. Another commonly used grouping defines EFD <0.08 mJ/mm^2 as low energy, medium energy as 0.08 to less than 0.28 mJ/mm^2, and high energy as 0.28–0.6 mJ/mm^2 [4].

Focused shockwave systems differ in their design and depend on whether the shockwaves are generated by electrohydraulic, electromagnetic, or piezoelectric mechanisms. The proposed mechanisms for the benefit of focused-ESWT include direct effects on tissue calcification, alteration of cell activity through cavitation, blocking the gate control mechanism, alteration for cell membrane permeability, enhanced angiogenesis, acoustic microstreaming, antiinflammatory effect, and effects on nociceptors through hyperstimulation [1–6].

Calcific or Non-Calcific Tendinopathy of the Rotator Cuff

Calcific Tendinopathy of the Rotator Cuff
Rotator cuff (RC) calcifications are a relatively common disease of the shoulder characterized by the presence of calcium hydroxyapatite crystal deposition in the tendons. The most common site of calcium deposit is at the supraspinatus tendon (80%), followed by infraspinatus, teres minor, and subscapularis tendon [7]. Diagnosis is made through the clinical presentations and imaging studies. Ultrasound is the most effective, sensitive, and inexpensive. MRI can rule out associated pathologies.

The natural history can evolve to spontaneous resolution over a period of time; however, the cycle can stagnate at any stage. Dr. Depalma and Kruper [8] pointed out that in some cases this initial improvement deteriorated and became a chronic carrier similar to patients with subacromial impingement symptoms. In this situation, more invasive procedures are indicated. Treatment depends on the intensity of symptoms, response to previous management, and the status of the evolving stage especially in Gärtner stages I and II.

The initial treatment is conservative, including rest, rehabilitation, analgesics, nonsteroidal anti-inflammatory drugs (NSAIDs), and subacromial injections, with favorable results in 90–99% of cases [7, 9, 10].

Dr. Gschwend et al. [11] states that the indications for invasive procedures are symptomatic progression, constant and intractable pain, and failure of conservative

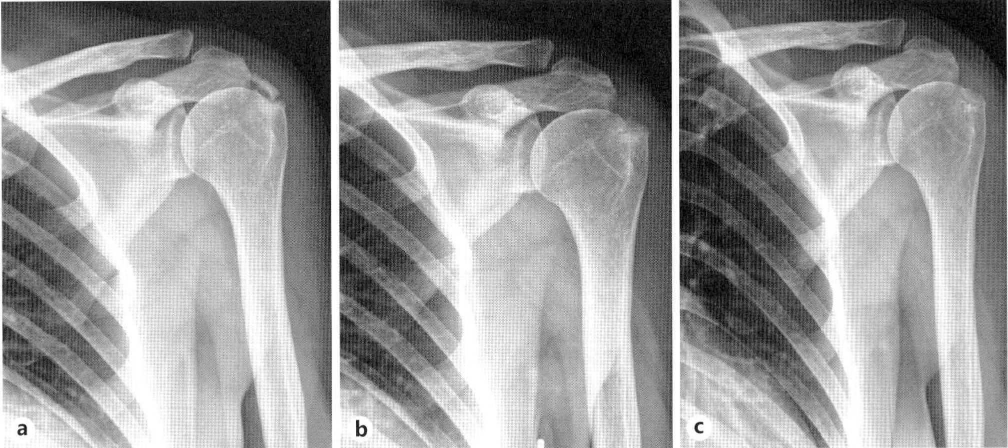

Fig. 1. a–c A 58-year-old female with calcific tedinopathy of the left shoulder.

treatment. In this situation, one can choose injection under ultrasound guidance or surgical treatment, through either open or arthroscopic method. Extracorporeal SW therapy (ESWT) has emerged as an alternative option when conservative treatment fails and prior to invasive procedures (Fig. 1a–c).

The mechanism of calcium absorption with ESWT has not been fully elucidated. Dr. Brañes et al. [12] demonstrated a neo-lymphangiogenesis phenomenon from biopsies done on samples taken from RC repair surgery, previously treated with ESWT.

Clinical-radiological dissociation is not uncommon. The persistence of calcification may be associated with a good clinical outcome. However, complete resorption reportedly has better results than partial resorption or persistence of the calcification.

Usually the ways to target the shockwaves on the calcification include topographic anatomy landmarks, feedback tender points from patient palpation, ultrasound, and radiology. Localization with fluoroscopy or computed tomography has reported to be more effective than when it was performed on the distal area of the supraspinatus tendon [13–16].

Although some studies showed that low energy could have satisfactory results, most publications showed a high-level energy to be more effective in calcification resorption and functional outcome [7, 17–23].

In a prospective comparative study, Dr. Wang et al. [24] showed that the ESWT group had 90.9% excellent or good results, with complete disappearance of calcification in 57.6% of patients, while the placebo group showed 83.3% poor results, and disappearance of calcification in 16.7% of patients.

In a randomized clinical trial of 144 patients, Dr. Gerdesmeyer et al. [7] reported better results in patients treated with either low- or high-energy ESWT, compared to those treated with placebo. Dr. Rompe et al. [25] compared the results of surgery with

ESWT and reported no difference in the functional outcomes at 1 year, and better improvement in patients treated with ESWT at 2 years.

After systematic review, Dr. Rebuzzi et al. [9] and Dr. Louwerens [26] suggested ESWT as the first therapeutic option because of its non-invasion. Absence of a dense calcification rim after ESWT is a good predictor of treatment outcome [9, 25–27]. Although evidence on the effectiveness of ESWT for calcific tendinopathy is adequate, there is no consensus on the most appropriate ESWT generator, energy level, frequency, number of sessions, number of impulses, use of anesthesia, or method of localization, which need further studies for reaffirmation [28]. The most commonly reported complications are local pain, intolerance, local petechiae, erythema, and hematomas. Another possible side effect is pain exacerbation, probably due to increased pressure within the subacromial space after ESWT treatment. Dr. Thiele [29] studied a series of 1,800 patients with calcific tendinopathy and found no major complications after 5 years of follow-up. The worst outcome included lack of an adequate response, with no definite clinical or radiographic improvement.

There are 2 studies reporting serious complications of humeral head necrosis after ESWT therapy. In one case, a 59-year patient developed cephalic necrosis 3 years after ESWT [30]. In the other case, necrosis appeared 3 months after therapy [31]. It has been discussed whether the initial painful symptoms were due to an early stage necrosis in these patients, and calcification could have occurred in a subclinical phase. Vascular lesions have been reported in patients undergoing renal lithotripsy, which might explain the mechanism of necrosis in these patients who had atypical vascularization of the proximal humerus. ESWT candidate patients should be informed of this rare but serious complication.

To sum up, ESWT is an efficacious and efficient alternative to surgery for treating RC calcification. ESWT treatment is efficient in pain and function improvement, and in resorption of calcification. It is safe, noninvasive, and avoids potential complications; it also reduces the recovery time and costs of surgery.

Nom-Calcific Tedinopathy of the RC
Non-calcific tendinopathy of the RC presents extrinsic and intrinsic pathogenic mechanisms. We prospectively reported the etiology of RC lesions and demonstrated the pathologic changes in the acromion and RC and the surgical results of patients with partial tears of the RC [32]. The degenerative grade of the RC on MRI was based on the criteria of Dr. Zlatkin et al. [33] and Dr. Iannotti. The pathology of the anterior acromion was graded using a modified system developed by Dr. Ozaki et al. [34] and Dr. Panni et al. [35] (Fig. 2).

MRI of the RC has shown more severe pathologic changes in articular side tears than bursal side tears. On the contrary, bursal side tears have more severe histological changes in the acromion. The clinical vignettes indicate that articular side tears of the RC mainly associate with intrinsic degenerative changes, whereas bursal side tears are mainly caused by subacromial impingement on the underlying more mild-

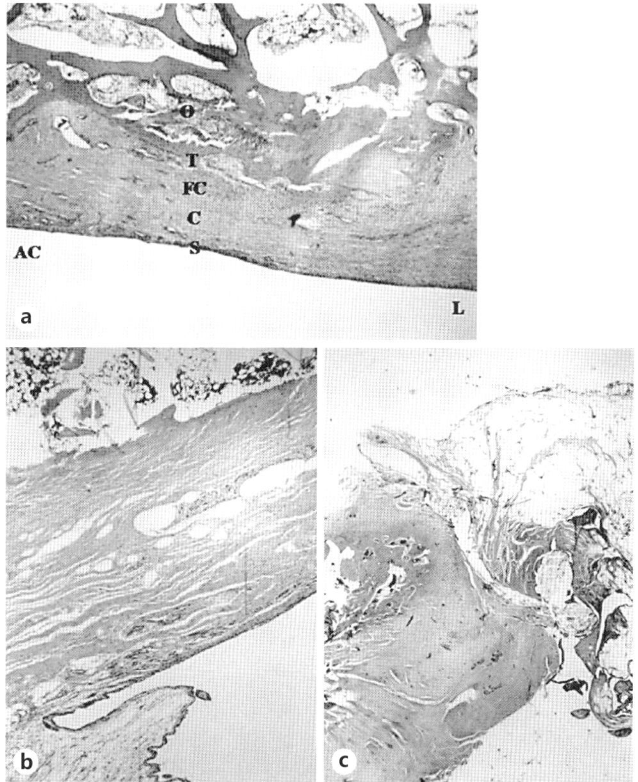

Fig. 2. Light micrographs of the anterior acromion showing different pathologic gradings for rotator cuff lesions. **a** Normal histologic features (grade 0) on the lateral side of the acromion (L) and a grade 1 degenerative pattern on the acromioclavicular joint side of the acromion (AC) are shown. The histologic features of the anterior acromion are well observed. O, osseous; T, tidemark; FC, fibrocartilaginous; C, collagenous; S, synovial layers (hematoxylin and eosin; original magnification 100x). **b** Grade 2 pathology shows a moderate degenerative pattern in the acromion (hematoxylin and eosin; magnification 100×). **c** A light micrograph of the anterior acromion shows a severe grade 3 degenerative pattern (hematoxylin and eosin; original magnification 100×).

ly degenerated RC. Therefore, management of both extrinsic subacromial impingement and intrinsic degeneration of the RC is recommended in the treatment of RC lesions.

ESWT cannot modify extrinsic outlet impingement but may improve vascularization of the RC and stimulate the release of growth factors [33]. In this situation, we could expect better vascularization and improved healing of the injured tissue, according to the histological results reported in treated tendons [12].

The clinical efficacy of ESWT in non-calcific RC tendinopathy is controversial [4, 21, 36–38]. Some authors showed that ESWT was not effective but did not clarify the exact etiology and pathogenesis in these patients [38]. On the other hand, some authors reported good results that analyzed the effect of ESWT in patients with subacromial impingement syndromes. ESWT and supervised exercise programs are complementary and not mutually exclusive; they need to be implemented together.

We thus believe that ESWT could have a complementary role in the treatment of chronic RC tendinopathy.

Although the available data do not allow conclusions on the effectiveness of ESWT in shoulder stiffness, research will continue with a larger series of patients to answer this difficult issue.

Summary
ESWT has emerged as strong therapeutic modality for shoulder pathology, especially for RC disorders. While many high-quality publications support its efficiency and efficacy in the treatment of RC calcifications, the results of ESWT in non-calcific RC tendinopathies remain controversial and need further research.

Patellar Tendinopathy (Jumper's Knee)

Patellar tendinopathy (PT), also described as "jumper's knee," is a common sport injury that happens in games involving jumping, running, and sudden changes of direction [39]. The overall prevalence of PT has been reported to be 14.2% in the general population, 44.6% in elite volleyball players, and 31.9% in basketball players [40].

The first stage of PT is described as acute reactive tendinopathy and early tendon disrepair phase, which in many cases progress to a second stage of chronic late disrepair and degeneration phase of the tendon [41].Even though PT has been related to high-demand physical activity, it may also occur in sedentary individuals. Some intrinsic risk factors of PT include overweight limb length discrepancy, hamstring muscle retractions, quadriceps stiffness, flat feet, or limited motion of the ankle joint [42–47]. The most relevant extrinsic risk factor and probably the major causes of PT are inappropriate training and exercise [43, 48, 49]. Sports gesture training including jumping and landing techniques, acceleration and deceleration protocols are crucial in preventing PT [50–52].

Chronic tendinopathy is an overuse syndrome manifested with pain and tenderness. The histology shows mucoid and chondroid degeneration, increased fibroblastic and myofibroblastic cells, and absent inflammatory cells [53]. Some studies showed that chronic painful tendinopathy exhibited increased nonvascular sensory, substance P-positive nerve fibers, and decreased vascular sympathetic nerve fibers. These suggested that the altered sensory-sympathetic innervation may play a role in the pathogenesis of tendinopathy [54].

The treatment of chronic PT is a big challenge. The patellar tendon is a non-stretching structure, which needs to deal with the largest muscle groups, the largest tensional loading forces, the longest bones, and the largest sesamoid in the human body. Biomechanically it is the most important point of the knee extensor mechanism, and is responsible for the function of jumping, running, and deceleration [55]. The vascular supply in the proximal insertion of patellar tendon is relatively poor; this may explain why tendon healing is poor on this site.

Many treatment strategies have been proposed for chronic PT, but there is no consensus on the most appropriate one [56–58]. Physical therapy, including eccentric exercise has shown the best evidence of good results [57–59]. Other types of conservative management include manual therapy, low-intensity pulsed ultrasound, and heavy-slow resistance training [60–63]. Injections of sclerosing agents or steroids also have good short-term results but have dose-related complications [63–66]. Biologic augmentation with autologous growth factors, platelet-rich plasma, bone marrow aspirates, or stem cells seem to be promising, but still lack evidence and have conflicting results [67–69].

Surgery probably is the last option in the treatment of chronic PT with good results in short and long terms. Open tenotomies with resection of dysplastic tissue, excision of Hoffa's fat pad, and drilling of the patella meet the main goal of the procedure: removal of angiofibroblastic tissue, revascularization, and pain control [70]. Even though the surgical complications are minimal, it is an invasive procedure with all the inconveniences and costs of a surgical protocol.

Understanding that chronic PT is caused by hypovascularity and fibrous dysplasia that changes the biomechanical properties of the tendon and causes pain, mechano-transduction stimulation of chronic tendinopathies has become a popular therapeutic tool with high effectiveness and safety [53, 71].

It is believed that shockwave therapy alleviates pain due to insertional tendinopathy by creating a loop of hyper stimulation analgesia, inducing neovascularization and improving blood supply to the tissue, and promoting repairs of the chronically inflamed tissues through tissue regeneration. Some authors have proven that ESWT decreases interleukins and tendon matrix metalloproteinase [53, 72–74]. Solid basic research has proved the biological responses that emanate through the cellular stimulation of metabolism and intercellular messengers via mechanotransduction [72].

Peers published a comparison between ESWT and surgery for PT [75]. He did not find any statistically significant differences in function or pain after 24 months. He concluded that ESWT and surgery are similar in results, but surgery had a higher cost, risk, recovery time, and inconveniences.

ESWT was also utilized in patients with PT secondary to harvesting of the patellar tendon for ACL reconstruction. Wang et al. [53] compared 30 knees in 27 patients treated with ESWT with 24 knees in 23 patients treated conservatively. At 2–3 year follow-up, the ESWT group had better outcome than control group [53].

Conclusion

Shockwave treatment for chronic patellar tendinopathies not responding to physical therapy and conservative measures is an effective and safe procedure. It is usually not recommended for acute conditions because more than 80% cases improve with physical therapy, and only 20% develop a dysplastic insertional tendinopathy. ESWT must be seen as part of treatment modalities for PT and not as an isolated treatment. The best results can be obtained when used in combination with standardized physical therapy protocols. Anesthesia is not recommended during the sessions. ESWT provides a safe, effective, and noninvasive alternative in the treatment of this real challenge in orthopaedics and sports medicine.

Achilles Tendinopathy

Chronic Achilles tendinopathy has been described as one of the most common overuse injuries in sports medicine [76, 77]. Recently, trends have clarified the difference between Achilles tendinitis and Achilles tendinosis. Achilles tendinitis (tendonitis) indicates local neuroinflammation processes and a clinical presence of pain and swelling [78–82]. Achilles tendinosis refers to a degenerative process without clinical or histologic signs of intratendinous inflammation. Instead, tendinosis is suggested to be caused by an imbalance between the degeneration and matrix synthesis [77, 79, 80, 83]. Patients with Achilles tendinopathy often complain of tendon pain by string exercises or weight-laded movements. Achilles tendinopathy can be classified as insertional, which occurs at the bone-tendon junction, and noninsertional, which occurs more proximally [79, 80, 82]. Insertional tendinopathy usually occurs in more active persons, whereas noninsertional tendinopathy tends to occur in overweight, less athletic, and older persons.

Common causes include gradual wear and tear from overuse, repetitive movements, or aging. Patients can also be injuried by simply damaging the tendon due to trauma. Intrinsic risk factors for Achilles tendinopathy include obesity, diabetes, aging, endocrinopathy, hyperpronation, forefoot varus deformity, limited mobility of the subtalar joint, and leg length discrepancy, Extrinsic risk factors include poor technique, mechanical overload, and training errors [77, 79, 80, 83–85].

Several treatment modalities including activity modification, arch supports, heel lifts, stretching exercises, eccentric loading, and nonsteroidal anti-inflammatory are recommended [81, 82, 86–89]. When conservative therapy fails, invasive treatment may be considered. The treatment of tendinopathies by ESWT has emerged as an alternative option, if conservative treatment fails prior to surgical interventions.

Dr. John Furia [90] conducted a case-controlled trial on non-insertional Achilles tendinopathy. He analyzed whether high-energy ESWT can be an effective treatment for chronic non-insertional Achilles tendinopathy or not, and recommended ESWT as an excellent option in chronic noninsertional Achilles tendinopathy.

Dr. Rasmussen et al. [91] also performed a double-blind randomized clinical trial for chronic Achilles tendinopathy. The authors concluded that ESET was an excellent option for chronic Achilles tendinopathy.

Dr. Rompe et al. [92] first conducted a study using radial shockwave therapy in the treatment of Achilles tendinopathy. The authors concluded that eccentric training or ESWT could be a good alternative treatment for patients with chronic Achilles tendinopathy.

Conclusion

Recently high-quality controlled trials have confirmed excellent results of ESWT in treating chronic Achilles tendinopathy. Further studies have to focus on the efficacy of different treatment modalities such as energy flux density or frequency and the application pressure. Shockwave therapy may be an effective option in the treatment of chronic Achilles tendinopathy.

Proximal Plantar Fasciitis and Lateral Epicondylitis of the Elbow

Proximal Plantar Fasciitis

Proximal plantar fasciitis (PPF) is usually diagnosed clinically based on the occurrence of morning heel pain made worse with ambulation and by the physical findings of tenderness over the medial aspect of the proximal plantar fascia. PPF has a bimodal distribution, afflicting both athletes and the sedentary. Imaging studies such as plain X-ray, sonography, CT or MRI, while generally not needed, can be helpful for ruling out other possibilities of heel pain or to establish the diagnosis of PPF when in doubt [93–98].

Initial treatment for PPT is non-operative and consists of rest, physical therapy, stretching, exercises, night splints, shoe inserts/orthotics, NSAIDs, and local injections. Patients who do not respond to conservative treatment for 4–6 months (between 10 and 20% of all patients) are candidates for more aggressive managements such as ESWT and surgery [93, 94, 99].

Many studies investigated the effect of ESWT in the treatment of PPF and reported a success rate ranging from 34 to 88% [76, 99–110]. The majority of these published papers reported a positive and beneficial effect of ESWT in the treatment of PPF.

Some studies compared the effect of ESWT with surgery, local corticosteroid injection, or physical therapy in the treatment of PPF [110–112]. ESWT and surgical treatment by plantar fasciotomy showed comparable functional outcomes. However, ESWT incurred no surgical risks, surgical pain, and complications [110]. Physical therapy has been shown to be comparable or better than ESWT in the treatment of PPF. However, physical therapy is time consuming and inconvenient [113]. Corticosteroid injection shows better short-term effects; however, there are dose-related complications and the long-term results favor ESWT [112].

The application of ESWT in the treatment of PPF is performed with either local anesthesia or no anesthesia. Some reports show that ESWT is less effective when the application is performed under local anesthesia [114, 115].

The complications of ESWT in the treatment of PPF are relatively low and negligible; they mainly include local reddening, ecchymosis, or mild hematoma, and migraine. These complications can be successfully managed conservatively and spontaneous recovery is anticipated.

In summary, the literature review unveiled controversy and discrepancy in the effect of ESWT on PPF. The vast majority of the published papers are in favor of ESWT.

Lateral Epicondylitis of the Elbow

Lateral epicondylitis of the elbow is a common upper extremity disorder with a reported incidence of 1–3% in the general population. The disorder is characterized by degenerative changes in the musculotendonous junction of the lateral epicondyle. It is usually caused by repetitive stress of flexion and extension movements of the wrist joint [116]. The disease is thought to be self-limiting, but it can cause severe pain and has a tendency to recur even after adequate treatment. There are various treatment options for the condition including resting, orthosis physical therapy, NSAID, and steroid injection. However, the optimal treatment method still remains elusive. Surgical managements are sometimes required when conservative treatments have failed [117, 118].

Several studies investigated the effect of ESWT in the treatment of lateral epicondylitis of the elbow, and the success rate ranged from 68 to 91% [119–123]. Dr. Park et al. [124] compared the therapeutic effect of ESWT for calcific and noncalcific lateral epicondylitis. They found that the therapeutic effect of ESWT for calcific lateral epicondylitis was not significantly different from that for noncalcific lateral epicondylitis. However, when a tendon tear exists, patients with calcific lateral epicondylitis might show poor prognosis after ESWT compared to patients with noncalcific lateral epicondylitis.

Dr. Lee et al. [125] evaluated the effectiveness of initial ESWT on the newly diagnosed lateral epicondylitis of the elbow and found that ESWT improved as much as the local steroid injection in treatment for lateral epicondylitis of the elbow.

References

1 Auge BK, Preminger GM: Update on shock wave lithotripsy technology. Curr Opin Urol 2002;12: 287–290.

2 Wess O UF, Durhben RN, et al: Working group technical developments – consensus report in high energy shock waves in medicine; in Chaussy C, Eisenberger F, Jocham D, et al. (eds): High Energy Shock Waves in Medicine. Thieme Suttgart, 1997.

3 Siebert W: Orthopädische Klinik Kassel, auf dem süddeutschen Orthopädenkongress, Baden Baden, 1996.

4 Bannuru RR, Flavin NE, Vaysbrot E, Harvey W, McAlindon T: High-energy extracorporeal shockwave therapy for treating chronic calcific tendinitis of the shoulder: a systematic review. Ann Intern Med 2014;160:542–549.

5 Speed CA: Extracorporeal shock-wave therapy in the management of chronic soft-tissue conditions. J Bone Joint Surg Br 2004;86:165–171.

6 Gleitz M: Myofascial Syndromes and Trigger Points – Shock Wave Therapy in Practice, ed 1. Heilbronn, Level 10 Publishing House Daniela Bamberg, 2011.

7 Gerdesmeyer L, Wagenpfeil S, Haake M, Maier M, Loew M, Wortler K, Lampe R, Seil R, Handle G, Gassel S, Rompe JD: Extracorporeal shock wave therapy for the treatment of chronic calcifying tendonitis of the rotator cuff: a randomized controlled trial. JAMA 2003;290:2573–2580.

8 Depalma AF, Kruper JS: Long-term study of shoulder joints afflicted with and treated for calcific tendinitis. Clin Orthop 1961;20:61–72.

9 Rebuzzi E, Coletti N, Schiavetti S, Giusto F: Arthroscopy surgery versus shock wave therapy for chronic calcifying tendinitis of the shoulder. J Orthop Traumatol 2008;9:179–185.

10 Suzuki K, Potts A, Anakwenze O, Singh A: Calcific tendinitis of the rotator cuff: Management options. J Am Acad Orthop Surg 2014;22:707–717.

11 Gschwend N, Patte D, Zippel J: [Therapy of calcific tendinitis of the shoulder]. Arch Orthop Unfallchir 1972;73:120–135.

12 Brañes J, Contreras HR, Cabello P, Antonic V, Guiloff LJ, Brañes M: Shoulder rotator cuff responses to extracorporeal shockwave therapy: Morphological and immunohistochemical analysis. Shoulder Elbow 2012;4:163–168.

13 Haake M, Deike B, Thon A, Schmitt J: Exact focusing of extracorporeal shock wave therapy for calcifying tendinopathy. Clin Orthop Relat Res 2002;397:323–331.

14 Sabeti M, Dorotka R, Goll A, Gruber M, Schatz KD: A comparison of two different treatments with navigated extracorporeal shock-wave therapy for calcifying tendinitis - a randomized controlled trial. Wien Klin Wochenschr 2007;119:124–128.

15 Charrin JE, Noel ER: Shockwave therapy under ultrasonographic guidance in rotator cuff calcific tendinitis. Joint Bone Spine 2001;68:241–244.

16 Sabeti-Aschraf M, Dorotka R, Goll A, Trieb K: Extracorporeal shock wave therapy in the treatment of calcific tendinitis of the rotator cuff. Am J Sports Med 2005;33:1365–1368.

17 Albert JD, Meadeb J, Guggenbuhl P, Marin F, Benkalfate T, Thomazeau H, Chales G: High-energy extracorporeal shock-wave therapy for calcifying tendinitis of the rotator cuff: a randomised trial. J Bone Joint Surg Br 2007;89:335–341.

18 Daecke W, Kusnierczak D, Loew M: [Extracorporeal shockwave therapy (ESWT) in tendinosis calcarea of the rotator cuff. Long-term results and efficacy]. Orthopäde 2002;31:645–651.

19 Huisstede BM, Gebremariam L, van der Sande R, Hay EM, Koes BW: Evidence for effectiveness of extracorporal shock-wave therapy (ESWT) to treat calcific and non-calcific rotator cuff tendinosis–a systematic review. Man Ther 2011;16:419–433.

20 Ioppolo F, Tattoli M, Di Sante L, Attanasi C, Venditto T, Servidio M, Cacchio A, Santilli V: Extracorporeal shock-wave therapy for supraspinatus calcifying tendinitis: a randomized clinical trial comparing two different energy levels. Phys Ther 2012;92:1376–1385.

21 Speed C: A systematic review of shockwave therapies in soft tissue conditions: Focusing on the evidence. Br J Sports Med 2014;48:1538–1542.

22 Vavken P, Holinka J, Rompe JD, Dorotka R: Focused extracorporeal shock wave therapy in calcifying tendinitis of the shoulder: a meta-analysis. Sports Health 2009;1:137–144.

23 Verstraelen FU, In den Kleef NJ, Jansen L, Morrenhof JW: High-energy versus low-energy extracorporeal shock wave therapy for calcifying tendinitis of the shoulder: Which is superior? A meta-analysis. Clin Orthop Relat Res 2014;472:2816–2825.

24 Wang CJ, Yang KD, Wang FS, Chen HH, Wang JW: Shock wave therapy for calcific tendinitis of the shoulder: a prospective clinical study with two-year follow-up. Am J Sports Med 2003;31:425–430.

25 Rompe JD, Zoellner J, Nafe B: Shock wave therapy versus conventional surgery in the treatment of calcifying tendinitis of the shoulder. Clin Orthop Relat Res 2001;387:72–82.

26 Louwerens JK, Veltman ES, van Noort A, van den Bekerom MP: The effectiveness of high-energy extracorporeal shockwave therapy versus ultrasound-guided needling versus arthroscopic surgery in the management of chronic calcific rotator cuff tendinopathy: a systematic review. Arthroscopy 2016;32:165–175.

27 Maier M, Stabler A, Lienemann A, Kohler S, Feitenhansl A, Durr HR, Pfahler M, Refior HJ: Shockwave application in calcifying tendinitis of the shoulder–prediction of outcome by imaging. Arch Orthop Trauma Surg 2000;120:493–498.

28 Ioppolo F, Tattoli M, Di Sante L, Venditto T, Tognolo L, Delicata M, Rizzo RS, Di Tanna G, Santilli V: Clinical improvement and resorption of calcifications in calcific tendinitis of the shoulder after shock wave therapy at 6 months' follow-up: A systematic review and meta-analysis. Arch Phys Med Rehabil 2013;94:1699–1706.

29 Thiele R: Tratamiento de las tendinosis calcificadas de hombro mediante ondas de choque. Experiencia de 1,800 casos y seguimiento a 5 años: Bienal Bosque Ortopedia. Bogotá, Universidad del Bosque, 2004.

30 Durst HB, Blatter G, Kuster MS: Osteonecrosis of the humeral head after extracorporeal shock-wave lithotripsy. J Bone Joint Surg Br 2002;84:744–746.

31 Liu HM, Chao CM, Hsieh JY, Jiang CC: Humeral head osteonecrosis after extracorporeal shock-wave treatment for rotator cuff tendinopathy. A case report. J Bone Joint Surg Am 2006;88:1353–1356.

32 Ko JY, Huang CC, Chen WJ, Chen CE, Chen SH, Wang CJ: Pathogenesis of partial tear of the rotator cuff: A clinical and pathologic study. J Shoulder Elbow Surg 2006;15:271–278.

33 Zlatkin MB, Iannotti JP, Roberts MC, Esterhai JL, Dalinka MK, Kressel HY, Schwartz JS, Lenkinski RE: Rotator cuff tears: diagnostic performance of mr imaging. Radiology 1989;172:223–229.

34 Ozaki J, Fujimoto S, Nakagawa Y, Masuhara K, Tamai S: Tears of the rotator cuff of the shoulder associated with pathological changes in the acromion. A study in cadavera. J Bone Joint Surg Am 1988;70:1224–1230.

35 Panni AS, Milano G, Lucania L, Fabbriciani C, Logroscino CA: Histological analysis of the coracoacromial arch: Correlation between age-related changes and rotator cuff tears. Arthroscopy 1996;12:531–540.

36 Kolk A, Yang KG, Tamminga R, van der Hoeven H: Radial extracorporeal shock-wave therapy in patients with chronic rotator cuff tendinitis: a prospective randomised double-blind placebo-controlled multicentre trial. Bone Joint J 2013;95-B:1521–1526.

37 Schmitt J, Haake M, Tosch A, Hildebrand R, Deike B, Griss P: Low-energy extracorporeal shock-wave treatment (ESWT) for tendinitis of the supraspinatus. A prospective, randomised study. J Bone Joint Surg Br 2001;83:873–876.

38 Speed CA, Richards C, Nichols D, Burnet S, Wies JT, Humphreys H, Hazleman BL: Extracorporeal shockwave therapy for tendonitis of the rotator cuff. A double-blind, randomised, controlled trial. J Bone Joint Surg Br 2002;84:509–512.

39 Blazina ME, Kerlan RK, Jobe FW, Carter VS, Carlson GJ: Jumper's knee. Orthop Clin North Am 1973;4:665–678.

40 Lian OB, Engebretsen L, Bahr R: Prevalence of jumper's knee among elite athletes from different sports: A cross-sectional study. Am J Sports Med 2005;33:561–567.

41 Cook JL, Purdam CR: Is tendon pathology a continuum? A pathology model to explain the clinical presentation of load-induced tendinopathy. Br J Sports Med 2009;43:409–416.

42 van der Worp H, van Ark M, Roerink S, Pepping GJ, van den Akker-Scheek I, Zwerver J: Risk factors for patellar tendinopathy: a systematic review of the literature. Br J Sports Med 2011;45:446–452.

43 Visnes H, Bahr R: Training volume and body composition as risk factors for developing jumper's knee among young elite volleyball players. Scand J Med Sci Sports 2013;23:607–613.

44 Kujala UM, Osterman K, Kvist M, Aalto T, Friberg O: Factors predisposing to patellar chondropathy and patellar apicitis in athletes. Int Orthop 1986;10:195–200.

45 Witvrouw E, Bellemans J, Lysens R, Danneels L, Cambier D: Intrinsic risk factors for the development of patellar tendinitis in an athletic population. A two-year prospective study. Am J Sports Med 2001;29:190–195.

46 Crossley KM, Thancanamootoo K, Metcalf BR, Cook JL, Purdam CR, Warden SJ: Clinical features of patellar tendinopathy and their implications for rehabilitation. J Orthop Res 2007;25:1164–1175.

47 Malliaras P, Cook JL, Kent P: Reduced ankle dorsiflexion range may increase the risk of patellar tendon injury among volleyball players. J Sci Med Sport 2006;9:304–309.

48 Ferretti A: Epidemiology of jumper's knee. Sports Med 1986;3:289–295.

49 Lian O, Refsnes PE, Engebretsen L, Bahr R: Performance characteristics of volleyball players with patellar tendinopathy. Am J Sports Med 2003;31:408–413.

50 Richards DP, Ajemian SV, Wiley JP, Zernicke RF: Knee joint dynamics predict patellar tendinitis in elite volleyball players. Am J Sports Med 1996;24:676–683.

51 Edwards S, Steele JR, McGhee DE, Beattie S, Purdam C, Cook JL: Landing strategies of athletes with an asymptomatic patellar tendon abnormality. Med Sci Sports Exerc 2010;42:2072–2080.

52 Taunton KM, Taunton JE, Khan KM: Treatment of patellar tendinopathy with extracorporeal shock wave therapy. Br Columbia Med J 2003;45:500–507.

53 Wang CJ, Ko JY, Chan YS, Weng LH, Hsu SL: Extracorporeal shockwave for chronic patellar tendinopathy. Am J Sports Med 2007;35:972–978.

54 Lian O, Dahl J, Ackermann PW, Frihagen F, Engebretsen L, Bahr R: Pronociceptive and antinociceptive neuromediators in patellar tendinopathy. Am J Sports Med 2006;34:1801–1808.

55 Kulig K, Landel R, Chang YJ, Hannanvash N, Reischl SF, Song P, Bashford GR: Patellar tendon morphology in volleyball athletes with and without patellar tendinopathy. Scand J Med Sci Sports 2013;23:e81–e88.

56 Gaida JE, Cook J: Treatment options for patellar tendinopathy: critical review. Curr Sports Med Rep 2011;10:255–270.

57 Larsson ME, Kall I, Nilsson-Helander K: Treatment of patellar tendinopathy–a systematic review of randomized controlled trials. Knee Surg Sports Traumatol Arthrosc 2012;20:1632–1646.

58 Rodriguez-Merchan EC: The treatment of patellar tendinopathy. J Orthop Traumatol 2013;14:77–81.

59 Visnes H, Bahr R: The evolution of eccentric training as treatment for patellar tendinopathy (jumper's knee): a critical review of exercise programmes. Br J Sports Med 2007;41:217–223.

60 Pedrelli A, Stecco C, Day JA: Treating patellar tendinopathy with fascial manipulation. J Bodyw Mov Ther 2009;13:73–80.

61 Stasinopoulos D, Stasinopoulos I: Comparison of effects of exercise programme, pulsed ultrasound and transverse friction in the treatment of chronic patellar tendinopathy. Clin Rehabil 2004;18:347–352.

62 Warden SJ, Metcalf BR, Kiss ZS, Cook JL, Purdam CR, Bennell KL, Crossley KM: Low-intensity pulsed ultrasound for chronic patellar tendinopathy: a randomized, double-blind, placebo-controlled trial. Rheumatology (Oxford) 2008;47:467–471.

63 Kongsgaard M, Kovanen V, Aagaard P, Doessing S, Hansen P, Laursen AH, Kaldau NC, Kjaer M, Magnusson SP: Corticosteroid injections, eccentric decline squat training and heavy slow resistance training in patellar tendinopathy. Scand J Med Sci Sports 2009;19:790–802.

64 Alfredson H, Ohberg L: Neovascularisation in chronic painful patellar tendinosis–promising results after sclerosing neovessels outside the tendon challenge the need for surgery. Knee Surg Sports Traumatol Arthrosc 2005;13:74–80.

65 Willberg L, Sunding K, Forssblad M, Fahlstrom M, Alfredson H: Sclerosing polidocanol injections or arthroscopic shaving to treat patellar tendinopathy/jumper's knee? A randomised controlled study. Br J Sports Med 2011;45:411–415.

66 Hoksrud A, Ohberg L, Alfredson H, Bahr R: Ultrasound-guided sclerosis of neovessels in painful chronic patellar tendinopathy: a randomized controlled trial. Am J Sports Med 2006;34:1738–1746.

67 Filardo G, Kon E, Della Villa S, Vincentelli F, Fornasari PM, Marcacci M: Use of platelet-rich plasma for the treatment of refractory jumper's knee. Int Orthop 2009;34:909–915.

68 Filardo G, Kon E, Di Matteo B, Pelotti P, Di Martino A, Marcacci M: Platelet-rich plasma for the treatment of patellar tendinopathy: clinical and imaging findings at medium-term follow-up. Int Orthop 2013;37:1583–1589.

69 Pascual-Garrido C, Rolón A, Makino A: Treatment of chronic patellar tendinopathy with autologous bone marrow stem cells: a 5-year-followup. Stem Cells Int 2012;2012:953510.

70 Maffulli N, Oliva F, Maffulli G, King JB, Del Buono A: Surgery for unilateral and bilateral patellar tendinopathy: A seven year comparative study. Int Orthop 2014;38:1717–1722.

71 van Leeuwen MT, Zwerver J, van den Akker-Scheek I: Extracorporeal shockwave therapy for patellar tendinopathy: a review of the literature. Br J Sports Med 2009;43:163–168.

72 van der Worp H, van den Akker-Scheek I, van Schie H, Zwerver J: Eswt for tendinopathy: technology and clinical implications. Knee Surg Sports Traumatol Arthrosc 2013;21:1451–1458.

73 Ioppolo F, Rompe JD, Furia JP, Cacchio A: Clinical application of shock wave therapy (SWT) in musculoskeletal disorders. Eur J Phys Rehabil Med 2014;50:217–230.

74 Wang CJ: Extracorporeal shockwave therapy in musculoskeletal disorders. J Orthop Surg Res 2012;7:11.

75 Peers KH, Lysens RJ, Brys P, Bellemans J: Cross-sectional outcome analysis of athletes with chronic patellar tendinopathy treated surgically and by extracorporeal shock wave therapy. Clin J Sport Med 2003;13:79–83.

76 Gerdesmeyer L, Frey C, Vester J, Maier M, Weil L Jr, Weil L Sr, Russlies M, Stienstra J, Scurran B, Fedder K, Diehl P, Lohrer H, Henne M, Gollwitzer H: Radial extracorporeal shock wave therapy is safe and effective in the treatment of chronic recalcitrant plantar fasciitis: results of a confirmatory randomized placebo-controlled multicenter study. Am J Sports Med 2008;36:2100–2109.

77 Diehl P, Gollwitzer H, Schauwecker J, Tischer T, Gerdesmeyer L: Konservative Therapie der chronischen Enthesiopathien. Orthopäde 2014;43:183–193.

78 Zwiers R, Wiegerinck JI, van Dijk CN: Treatment of midportion achilles tendinopathy: an evidence-based overview. Knee Surg Sports Traumatol Arthrosc 2014;24:2103–2111.

79 Magnan B, Bondi M, Pierantoni S, Samaila E: The pathogenesis of achilles tendinopathy: a systematic review. Foot Ankle Surg 2014;20:154–159.

80 Joseph MF, Lillie KR, Bergeron DJ, Cota KC, Yoon JS, Kraemer WJ, Denegar CR: Achilles tendon biomechanics in response to acute intense exercise. J Strength Cond Res 2014;28:1181–1186.

81 Rowe V, Hemmings S, Barton C, Malliaras P, Maffulli N, Morrissey D: Conservative management of midportion achilles tendinopathy: a mixed methods study, integrating systematic review and clinical reasoning. Sports Med 2012;42:941–967.

82 Notarnicola A, Maccagnano G, Tafuri S, Fiore A, Margiotta C, Pesce V, Moretti B: Prognostic factors of extracorporeal shock wave therapy for tendinopathies. Musculoskelet Surg 2016;100:53–61.

83 Knobloch K: The role of tendon microcirculation in achilles and patellar tendinopathy. J Orthop Surg Res 2008;3:18.

84 Kearney R, Costa ML: Insertional achilles tendinopathy management: A systematic review. Foot Ankle Int 2010;31:689–694.

85 Furia JP: Extrakorporale Stosswellentherapie zur Behandlung der Achillessehnentendinopathie. Orthopäde 2005;34:571–578.

86 Childress MA, Beutler A: Management of chronic tendon injuries. Am Fam Physician 2013;87:486–490.

87 Wiegerinck JI, Kerkhoffs GM, van Sterkenburg MN, Sierevelt IN, van Dijk CN: Treatment for insertional achilles tendinopathy: a systematic review. Knee Surg Sports Traumatol Arthrosc 2013;21:1345–1355.

88 Balasubramaniam U, Dissanayake R, Annabell L: Efficacy of platelet-rich plasma injections in pain associated with chronic tendinopathy: a systematic review. Phys Sportsmed 2015;43:253–261.

89 Maffulli N, Papalia R, D'Adamio S, Diaz Balzani L, Denaro V: Pharmacological interventions for the treatment of achilles tendinopathy: a systematic review of randomized controlled trials. Br Med Bull 2015;113:101–115.

90 Furia JP: High-energy extracorporeal shock wave therapy as a treatment for chronic noninsertional achilles tendinopathy. Am J Sports Med 2008;36:502–508.

91 Rasmussen S, Christensen M, Mathiesen I, Simonson O: Shockwave therapy for chronic achilles tendinopathy: a double-blind, randomized clinical trial of efficacy. Acta Orthop 2008;79:249–256.

92 Rompe JD, Nafe B, Furia JP, Maffulli N: Eccentric loading, shock-wave treatment, or a wait-and-see policy for tendinopathy of the main body of tendo achillis: a randomized controlled trial. Am J Sports Med 2007;35:374–383.

93 Rompe JD: Plantar fasciopathy. Sports Med Arthrosc 2009;17:100–104.

94 Thomas JL, Christensen JC, Kravitz SR, Mendicino RW, Schuberth JM, Vanore JV, Weil LS Sr, Zlotoff HJ, Bouché R, Baker J; American College of Foot and Ankle Surgeons heel pain committee: The diagnosis and treatment of heel pain: A clinical practice guideline-revision 2010. J Foot Ankle Surg 2010;49(3 suppl):S1–S19.

95 Tong KB, Furia J: Economic burden of plantar fasciitis treatment in the United States. Am J Orthop (Belle Mead NJ) 2010;39:227–231.

96 Furia JP, Rompe JD: Extracorporeal shock wave therapy in the treatment of chronic plantar fasciitis and achilles tendinopathy. Curr Opin Orthop 2007; 18:101–111.

97 Rompe JD, Furia J, Weil L, Maffulli N: Shock wave therapy for chronic plantar fasciopathy. Br Med Bull 2007;81–82:183–208.

98 Neufeld SK, Cerrato R: Plantar fasciitis: Evaluation and treatment. J Am Acad Orthop Surg 2008;16: 338–346.

99 Buch M, Knorr U, Fleming L, Theodore G, Amendola A, Bachmann C, Zingas C, Siebert WE: [Extracorporeal shockwave therapy in symptomatic heel spurs. An overview]. Orthopäde 2002;31:637–644.

100 Chen HS, Chen LM, Huang TW: Treatment of painful heel syndrome with shock waves. Clin Orthop Relat Res 2001;387:41–46.

101 Chuckpaiwong B, Berkson EM, Theodore GH: Extracorporeal shock wave for chronic proximal plantar fasciitis: 225 patients with results and outcome predictors. J Foot Ankle Surg 2009;48:148–155.

102 Gollwitzer H, Diehl P, von Korff A, Rahlfs VW, Gerdesmeyer L: Extracorporeal shock wave therapy for chronic painful heel syndrome: A prospective, double blind, randomized trial assessing the efficacy of a new electromagnetic shock wave device. J Foot Ankle Surg 2007;46:348–357.

103 Ibrahim MI, Donatelli RA, Schmitz C, Hellman MA, Buxbaum F: Chronic plantar fasciitis treated with two sessions of radial extracorporeal shock wave therapy. Foot Ankle Int 2010;31:391–397.

104 Kudo P, Dainty K, Clarfield M, Coughlin L, Lavoie P, Lebrun C: Randomized, placebo-controlled, double-blind clinical trial evaluating the treatment of plantar fasciitis with an extracoporeal shockwave therapy (ESWT) device: a north American confirmatory study. J Orthop Res 2006;24:115–123.

105 Metzner G, Dohnalek C, Aigner E: High-energy extracorporeal shock-wave therapy (ESWT) for the treatment of chronic plantar fasciitis. Foot Ankle Int 2010;31:790–796.

106 Ogden JA, Alvarez RG, Levitt RL, Johnson JE, Marlow ME: Electrohydraulic high-energy shock-wave treatment for chronic plantar fasciitis. J Bone Joint Surg Am 2004;86-A:2216–2228.

107 Rompe JD, Decking J, Schoellner C, Nafe B: Shock wave application for chronic plantar fasciitis in running athletes. A prospective, randomized, placebo-controlled trial. Am J Sports Med 2003;31: 268–275.

108 Wang CJ, Chen HS, Chen WS, Chen LM: Treatment of painful heels using extracorporeal shock wave. J Formos Med Assoc 2000;99:580–583.

109 Wang CJ, Wang FS, Yang KD, Weng LH, Ko JY: Long-term results of extracorporeal shockwave treatment for plantar fasciitis. Am J Sports Med 2006;34:592–596.

110 Weil LS Jr, Roukis TS, Weil LS, Borrelli AH: Extracorporeal shock wave therapy for the treatment of chronic plantar fasciitis: Indications, protocol, intermediate results, and a comparison of results to fasciotomy. J Foot Ankle Surg 2002;41:166–172.

111 Othman AM, Ragab EM: Endoscopic plantar fasciotomy versus extracorporeal shock wave therapy for treatment of chronic plantar fasciitis. Arch Orthop Trauma Surg 2010;130:1343–1347.

112 Yucel I, Ozturan KE, Demiraran Y, Degirmenci E, Kaynak G: Comparison of high-dose extracorporeal shockwave therapy and intralesional corticosteroid injection in the treatment of plantar fasciitis. J Am Podiatr Med Assoc 2010;100:105–110.

113 Greve JM, Grecco MV, Santos-Silva PR: Comparison of radial shockwaves and conventional physiotherapy for treating plantar fasciitis. Clinics (Sao Paulo) 2009;64:97–103.

114 Labek G, Auersperg V, Ziernhöld M, Poulios N, Böhler N: Einfluss von Lokalanästhesie und Energieflussdichte bei niederenergetischer extrakorporaler Stosswellentherapie der chronischen plantaren Fasziitis. Z Orthop Ihre Grenzgeb 2005;143: 240–246.

115 Rompe J, Meurer A, Nafe B, Hofmann A, Gerdesmeyer L: Repetitive low-energy shock wave application without local anesthesia is more efficient than repetitive low-energy shock wave application with local anesthesia in the treatment of chronic plantar fasciitis. J Orthop Res 2005;23:931–941.

116 Kraushaar BS, Nirschl RP: Tendinosis of the elbow (tennis elbow). Clinical features and findings of histological, immunohistochemical, and electron microscopy studies. J Bone Joint Surg Am 1999;81: 259–278.

117 Boyd HB, McLeod AC Jr: Tennis elbow. J Bone Joint Surg Am 1973;55:1183–1187.

118 Friedlander HL, Reid RL, Cape RF: Tennis elbow. Clin Orthop Relat Res 1967;51:109–116.

119 Ko JY, Chen HS, Chen LM: Treatment of lateral epicondylitis of the elbow with shock waves. Clin Orthop Relat Res 2001;387:60–67.

120 Ozturan KE, Yucel I, Cakici H, Guven M, Sungur I: Autologous blood and corticosteroid injection and extracoporeal shock wave therapy in the treatment of lateral epicondylitis. Orthopedics 2010;33:84–91.

121 Radwan YA, ElSobhi G, Badawy WS, Reda A, Khalid S: Resistant tennis elbow: Shock-wave therapy versus percutaneous tenotomy. Int Orthop 2008; 32:671–677.

122 Rompe JD, Decking J, Schoellner C, Theis C: Repetitive low-energy shock wave treatment for chronic lateral epicondylitis in tennis players. Am J Sports Med 2004;32:734–743.

123 Spacca G, Necozione S, Cacchio A: Radial shock wave therapy for lateral epicondylitis: A prospective randomised controlled single-blind study. Eura Medicophys 2005;41:17–25.

124 Park JW, Hwang JH, Choi YS, Kim SJ: Correction: Comparison of therapeutic effect of extracorporeal shock wave in calcific versus noncalcific lateral epicondylopathy. Ann Rehabil Med 2016;40:557.

125 Lee SS, Kang S, Park NK, Lee CW, Song HS, Sohn MK, Cho KH, Kim JH: Effectiveness of initial extracorporeal shock wave therapy on the newly diagnosed lateral or medial epicondylitis. Ann Rehabil Med 2012;36:681–687.

Jih-Yang Ko, MD
Department of Orthopedic Surgery, Graduate Institute of Clinical Medical Sciences
Chang Gung University College of Medicine, Kaohsiung Chang Gung Memorial Hospital
123, Ta-Pei Road, Niao-Sung District
Kaohsiung 83303 (Taiwan)
E-Mail kojy@cgmh.org.tw

Wang C-J, Schaden W, Ko J-Y (eds): Shockwave Medicine.
Transl Res Biomed. Basel, Karger, 2018, vol 6, pp 42–63 (DOI: 10.1159/000485061)

Significance of Extracorporeal Shockwave Therapy in Fracture Treatment

Nicolas Haffner[a] · Daniel Smolen[a, b] · Falko Dahm[a, c] ·
Wolfgang Schaden[a, d, e] · Rainer Mittermayr[a, c, e, f]

[a]Ludwig Boltzmann Institute for Experimental and Clinical Traumatology, Vienna, Austria; [b]Etzel Clinic,
Pfäffikon, Switzerland; [c]AUVA Trauma Center Meidling, Vienna, [d]AUVA Medical Board, Vienna, and
[e]AUVA Trauma Research Center, Vienna, and [f]Austrian Cluster for Tissue Engineering, Vienna, Austria

Abstract

Numerous studies have proven that extracorporeal shockwave therapy (ESWT) is effective in the noninvasive treatment of delayed healing fractures and nonunions. Imposing shockwave therapy as first-line treatment for the above-mentioned indications could significantly reduce costs and surgery associated complications. Further applications comprising acute or fragility fractures prone to develop healing disorders as well as stress fractures also show promising results after ESWT. In contrast to the past, where the mechanistic (destructive) model as method of action of ESWT was favored, recent studies have brought light into the fundamental working mechanisms of ESWT. Shockwaves are converted in the target tissue into biochemical signals via activating different signaling pathways (mechanism called mechanotransduction), which reflect the stimulation of the endogenous regeneration potential. This chapter tries to summarize the recent literature of ESWT in the treatment of bone healing disturbances.
© 2018 S. Karger AG, Basel

Introduction

Fracture healing as being a complex physiological process, involving multiple steps including but not limited to angiogenesis and callus formation in order to allow osseous consolidation, is orchestrated by different proteins. Transcription and translation of these proteins is strictly tied to a predetermined time course. Bone has the unique ability to restore continuity and function (= regeneration) without scar tissue unlike other tissues, which usually end up being repaired. Repairs in general involve scars, which exhibit inferior mechanical properties than the original tissue.

However, under certain circumstances, physiological fracture healing may be disturbed, thus leading to impaired or delayed healing, which is the case in about 5–10% of the 5.6 million fractures that occur annually in the United States. It is estimated, that up to 10% of all fractures eventually need further surgical interventions for impaired healing [1].

Although nonunions serendipitously occur on rare occasions (approximately 2.5% of all fractures in Austria develop a nonunion), yet they represent a highly relevant medical topic and socioeconomic burden. Comparisons of direct and indirect costs between countries prove to be difficult due to vast discrepancies in treatment algorithms and reimbursement policies. Sprague and Bhandari [2] estimated direct and indirect costs of delayed healings in Canada to be around USD 80,000 per case.

Definition of nonunions remains to be controversial as is frequently debated in the literature. Clinicians apply different classifications in order to discriminate delayed- from non-healing fractures, which hinder comparison of outcomes.

Normal fracture healing in long bones usually occurs within 3 months. Therefore, a fracture failing to show osseous healing within this time period can be defined as delayed healing. Consequently, the absence of osseous consolidation beyond 6 months is often considered nonunion. The Federal Drug and Food Administration, however, defines nonunion as fracture that does not heal without additional interventions within 9 months, lacking any radiological signs of bone formation over the past 3 consecutive months [3].

Contributing factors, potentially yielding in delayed healing or nonunions can be divided into local and systemic. Local factors comprise the fracture type and anatomic region as well as soft-tissue destruction and its influence on vascularization associated with the primary trauma. Accordingly, the tibia among all long bones is more prone to develop healing disturbances due to its limited soft tissue envelope and the higher incidence rate of open and comminuted fractures.

Systemic factors include comorbidities such as diabetes, atherosclerosis, congestive heart disease, and lifestyle habits such as smoking and alcohol consumption, as well as current medication and patient's age.

Due to the fact that according to the Federal Drug and Food Administration definition nonunions lack the intrinsic potential to heal on their own, various surgical treatments have been clinically established in order to realize osseous consolidation. Techniques include but are not limited to intramedullary nailing (unreamed/reamed), compression plating to address fracture stability, and autologous bone grafting as supplementary osteoinductive and -conductive measures. In selected cases as in septic nonunions, external fixation systems (e.g., Illizarov technique) are a suitable choice of treatment [4]. However, surgical intervention inherently compromises vascularization, and hence potentially negatively influences healing, not to mention, the relatively high frequency of donor site morbidity after autologous bone graft harvesting (e.g., from the iliac crest) affecting patients' quality of life [5]. Thus, miscellaneous semi- or noninvasive alternatives were investigated to avoid complications associated

Table 1. Influence of the total amount of impulses/energy on healing of nonunions

Pulses	Number	Healed	Failure
<3,000	253	**183 (72)**	70 (28)
>3,000 <4,000	224	**177 (79)**	47 (21)
>4,000 <6,000	40	26 (65)	14 (35)
>6,000 <8,000	37	23 (62)	14 (38)
>8,000–12,000	93	65 (70)	28 (30)
Total	647	477 (73)	173 (27)

Figures in parentheses are percentages. Lower total amount of impulses yielded better outcome, consequently applying a lower amount of impulses as standard [18].

with surgery. However, low-intensity pulsed ultrasound [6–8] or the application of pulsed electromagnetic fields [9–11] as well as the use of osteogenic growth factors such as bone morphogenetic proteins [12–14] have shown either conflicting results or an irrational cost to benefit ratio.

Among noninvasive treatment methods, extracorporeal shockwave treatment (ESWT) has evolved as a valuable and reproducible alternative in bone pathologies. Initially intended to treat urinary concrements, Dr. Haupt et al. [15] incidentally observed cortical thickening (hypertrophy) of the iliac bone. Accordingly, ESWT was introduced to bone pathologies first and foremost to delayed healings and nonunions. Thereafter, Dr. Valchanou and Michailov [16] published already in 1991 their first data on nonunions, presuming that ESWT creates microfractures and consequently induces healing (repair due to destruction). This hypothesis remained to be the predominant mode of action of ESWT, until Dr. Tischer et al. [17] raised concerns after investigating histological specimens exposed to ESWT well below the energy flux densities usually used to treat urinary stones (urolithotripsy). The specimen revealed bone formation without signs of destruction or microseparation. Consistently, Dr. Schaden et al. [18] accidentally observed similar dose-dependent effects in the treatment of nonunions, as bony union was achieved even though only a third of the intended impulses were applied due to a technical failure of the device (Table 1). Since then, basic research provided new insights into the working mechanism of ESWT.

Dr. Wang et al. [19] showed in their experiments including immunohistochemical stainings that ESWT induces neovascularization/angiogenesis. Histology confirmed no destruction in these experiments. Another study by Dr. Wang et al. [20] demonstrated the upregulation and consequently the increased expression of angiogenic and osteogenic growth factors. Upregulation lasted for about 12 weeks. Later on, research focused more and more on biological effects of ESWT and hence the mechanistic (destructive) model which was first hypothesized to be the cause of effect in ESWT, was replaced by the biological model.

In recent studies, the term mechanotransduction was used in order to describe effects of ESWT on tissues [21]. Recent studies also confirm mechanotransduction as the mode of primary action of ESWT [22, 23].

Working Mechanism of ESWT in Bone Tissue and Fracture Repair

While numerous studies have already proven early the positive clinical effects of high-energy shockwaves in the treatment of nonunions, their fundamental working mechanism was only recently revealed. Owing to profound and extensive basic research activities in recent years, we can meanwhile draw a more comprehensive picture.

A shockwave is a physically well-characterized short-lasting acoustic impulse, which induces compressive- shear-, and tensile-forces at tissue interfaces. The energy of a shockwave rapidly dissipates with distance. The above-mentioned forces activate signaling cascades through direct and indirect mechanisms of action. This translation from the physical origin as acoustic impulse into biochemical signals is called mechanotransduction. Subsequently, the signal leads to gene expression and thereafter to transcription and translation finally ending in protein synthesis, resulting in a spatial and chronological concentrated healing process [24].

Dr. Xu et al. [23] demonstrated that ESWT promotes the expression of transmembranous integrins on the surface of osteoblasts, particularly or most notably $\alpha 1\beta 5$, an integrin that is actively involved in the interaction between the extracellular matrix (ECM) and the cell. This increased expression of integrins, significantly elevated phosphorylated focal adhesion kinase (FAK), which is supposed to be the key element of signal transduction pathways triggered by integrins. In vitro experiments on mesenchymal stem cells supported this finding of increased phosphorylation of FAK in response to shockwaves but through the inhibition of miR-138, known as a direct inhibitor of the FAK gene during osteogenic differentiation of bone marrow stem cells [25]. Another mechanism of delivering energy of ESWT from outside the cell to the cytoplasm was identified by changes in the transmembrane current due to increased K^+ and Ca^{2+} influx [26].

All these alterations on cell membranes and the ECM evoked by shockwaves stimulate a number of downstream intracellular signaling cascades.

The integrin-induced phosphorylation of FAK by ESWT was shown to induce the activation of ERK1/2 via MEK1/2, thus leading to an increased adhesion, distribution, and migration of osteoblasts, finally stimulating fracture healing [23]. In a bone defect model, the activation of ERK and p38/MAPK via shockwaves has been demonstrated to enhance the mitogenic cell activity dedicated for chondro- and osteogenesis [27]. Interestingly, ERK phosphorylation via Ras and Rac1 protein in osteoblasts was also seen by superoxide radical, which was elevated after shockwave treatment [28, 29]. Subsequent enhanced expression of HIF-1α and VEGF resulted in angiogenesis, the latter being already demonstrated for other cell and tissue types following ESWT. ERK activation following shockwave-mediated radical generation could also be verified in mesenchymal stem cells, which led to increased proliferation and differentiation into osteoprogenitor cells via the osteogenic transcription factor CBFA1. Consequently, augmented osteogenesis occurred through shockwave-induced ERK pathway-dependent activation of CBFA1 transcription factor [30]. On a protein level, the activation of the ERK signaling pathway increased RUNX-2, a major transcriptional factor for osteogenesis [25]. On the other hand, a decreased RANKL/OPG ratio was found in shockwave-stimulated osteoblasts, indicating inhibition of osteoclastogenesis [31]. Both, the increased RUNX-2 protein level and the decreased RANKL/OPG ratio favor bone repair. Affymetrix microarrays could even demonstrate an upregulation of numerous different genes not only involved in osteoblast differentiation but also in bone formation, skeletal development, and cell homeostasis [32]. Gene expression analysis of ECM proteins by in-situ hybridization in shockwave revealed that rodent femurs exhibited a spatial and temporal regulation of osteogenic cells. Upregulation of pro-a 1 (I) collagen, osteocalcin, and osteopontin was detected in subperiosteal osteoblastic cells 4 days thereafter, thus leading to periosteal bone formation. This enhanced gene regulation lasted up to day 14, but in different amounts and partially with a different cellular origin. After 3 weeks, the shockwave-induced osteogenic stimulus resulted in elevated bone mineral content and bone mineral density compared to the internal contralateral femur controls [33]. In vitro studies on periosteal cells could also demonstrate that shockwave treatment has a stimulatory effect on these cells, but in a delayed pattern, contrary to the study conducted by Takahashi. While on day 6 after exposure of human periosteal cells to shockwaves, the activity of alkaline phosphatase was in fact decreased compared to controls, it was not until day 18 that cell proliferation and an elevated ALP content could be measured. Consequently, on day 35, an increased mineralization of ECM was evident [34, 35]. However, not only mineralization of the ECM is stimulated by ESWT, but also sulfated glycosaminoglycans, as main matrix components, were increasingly observed over the course of 28 days after shockwave exposure [36].

Acoustic tissue stimulation by shockwaves has further demonstrated to produce and liberate various growth factors involved in bone repair including but not limited to bone morphogenetic proteins, TGF-1, and IL-10 although the specific pathway was not further elucidated [27, 37–41].

The working mechanism of ESWT was not only shown for bone tissue and in fracture repair but also in different other cell types and tissues. In this regard, shockwave activated intracellular signaling cascades including p38, MAPK, ERK1/2 [22, 42, 43], and AKT [44–46]. Additionally, it was shown that ESWT influences or induces the recruitment of stem cells partially via SDF-1 [47–49]. Dr. Schuh et al. [50] observed a prolonged expression and an overall increase in mesenchymal stem cell markers after ESWT. Consistently first applications of combined therapies, including ESWT and autologous stem cell transplantation in patients affected by nonunions, yielded an increase in their healing rate [51]. Furthermore, in a tissue-engineering approach, suc-

  Haffner · Smolen · Dahm · Schaden · Mittermayr

cessful orthotopic bone formation could be achieved by seeding a bone scaffold with shockwave-stimulated periosteal cells [52].

In summary, ESWT interacts on multiple cellular and molecular levels, inducing regeneration rather than repair.

ESWT in the Treatment of Acute Fractures

Breaking a bone is a dramatic experience for every patient. Besides causing pain and loosing function of the affected bone, the fracture itself potentially implies high direct and indirect socioeconomic costs. Successful treatment of acute fractures includes conservative and surgical measures. Irrespective of the initial treatment, whether it was conservative or surgical, a distinct after-care is important. However, healing can be profoundly disturbed yielding in delayed or non-healing fractures due to local and systemic influencing variables. Factors potentially disturbing physiological healing, which can be well recognized at initial evaluation include the severity of comminution, the fracture-associated extent of soft tissue injury, and comorbidities such as diabetes and vascular disease. The surgical approach (technique as well as type of osteosynthesis) in these cases is of paramount importance in order to avoid or at least to reduce the incidence of healing failures. Consistently extensive operative approaches in order to establish anatomic reduction and absolute stability have been replaced more and more by biologic approaches such as minimal invasive plate osteosynthesis. The key to these minimal-invasive approaches is the fact that periosteal stripping as well as further damage to the soft-tissue envelope is avoided. Seeking to prevent complications, namely, delayed healings and nonunions and according to slogans such as "the more the better" and "better safe than sorry," adjunctive methods to surgery have always been debated upon.

In this respect, ESWT represents a valuable prophylactic or adjunctive tool in operative treatments of acute fractures, as it can be applied simultaneously without a noteworthy prolongation of OR time. Dr. Wang et al. [53] was the first to investigate the application of ESWT in acute high-energy fractures. In this prospective randomized study, the authors included 56 patients with 59 acute long bone fractures of the lower extremity. The study and control group were homogenously distributed in demographic patients' data, fracture location, operative procedure, and aftercare treatment, leaving ESWT as the only difference between both groups. ESWT was applied using an electrohydraulic device at the end of the surgical procedure, delivering 6,000 impulses at an energy flux density of 0.62 mJ/mm^2. Radiographic examinations revealed a significantly higher union rate in the study group which was most evident at 6-month follow-up (63 vs. 23% in the control group). At 12 months, the difference was still significant (89 vs. 80%). Consistently, they concluded that ESWT as adjunctive measure leads to a faster and higher union rate. Similarly, Dr. Moretti et al. [54] demonstrated in their case series of 16 acute closed tibial fractures an accelerated healing depicted as a higher number of healed cortices. Independent radiological analysis revealed cortical continuity with an average of 3.25 in the study group compared to 2.54 in the control group.

Supporting the hypothesis that ESWT enhances bone healing in acute fractures, Dr. Kieves et al. [55] studied the influence of shockwaves in a canine tibial osteotomy model. Assuming osteotomy as an acute iatrogenic fracture, their study revealed a significantly better outcome after 8 weeks in terms of fracture healing in the treatment group. Both groups underwent a leveling tibial plateau osteotomy followed by plate fixation. In contrast to Dr. Wang [53], ESWT was applied twice (intraoperatively after plate fixation and 2 weeks thereafter), introducing 1,000 impulses/session with an electrohydraulic device at an energy flux density of 0.15 mJ/mm^2.

Although these studies suggest a positive influence of ESWT, further RCTs are needed to elucidate the adjunctive value of ESWT in the setting of acute fracture treatment. In order to include ESWT or any other adjunctive method to surgery or even to overall treatment of acute fractures as standard of care, it must prove superior to surgery or conservative treatment alone, preferably minimizing or even eliminating additional harms and costs for the patient.

Currently, no established treatment modalities are available to support healing in fractures, especially in those prone to develop delayed healings or nonunions. Nonetheless, different other biophysical means are under investigation to countervail such complications including low-intensity pulsed ultrasound and pulsed electromagnetic fields. However, focused high-energy ESWT has shown the potential to positively influence bone healing and above all, the electrohydraulic device might be the most applicable and cost-effective measure due to its single application.

ESWT in Delayed Healing Fractures and Nonunions

Over the past 2 decades, several studies showed successful application of ESWT in the treatment of delayed or non-healing fractures [56–59]. Due to our own experience with a case series of 115 consecutive patients, published in 1998, achieving bony healing after 6 months following ESWT in 87 nonunions (75.7% union rate), ESWT was established as first-line treatment for nonunions at our institution [18]. Since then, over 3,500 delayed or non-healing fractures at different anatomic locations were treated with an average success rate of almost 80% after 6-months follow-up (data not published). The rationale of imposing ESWT as standard of care or fist-line treatment of nonunions in our institution included not only the high success rate but also the almost complete absence of complications comparing favorably to the so-called gold standard being revision surgery. Besides the clear advantages for ESWT and most notably the preference of most patients not to undergo major surgery with the associated risks and complications, financial considerations have also led to increasing recognition of ESWT worldwide.

According to the requirements of evidence-based medicine that a treatment modality has to be at least equally effective if not superior as the standard of care, Dr. Cacchio et al. [60] conducted a prospective, randomized, controlled, multicenter trial (evidence level I) investigating treatment effects of ESWT in long bone nonunions (radius, ulna, femur, and tibia) compared to "standard-of-care" surgery. In order to eliminate bias, study groups were meticulously matched and homogenously distributed. ESWT was performed in two modalities differing in their energy flux density being 0.4 mJ/mm^2 in group 1 and 0.7 mJ/mm^2 in group 2. Both groups received 4,000 impulses/session with an electromagnetic device focusing on the fracture gap. Treatment was administered 4 times in a weekly interval. The surgical control group followed a standard protocol in order to guarantee maximum reproducibility. The protocol included the removal of the implant, decortication, and debridement of the fracture site, reopening of the medullary canal, followed by fracture reduction and fixation, as well as autologous bone grafting if deemed necessary. Radiological evaluation, being the primary outcome parameter, was performed at various time points up to 24 months after intervention. The radiographic findings among study groups did not differ significantly and osseous healing was seen in 94% in group 1 (0.4 mJ/mm^2), in 92% in group 2 (0.7 mJ/mm^2), and in 95% in the surgical group after 2 years. However, the Visual Analogue Scale, the Lower Extremity Functional Scale as well as the Disabilities of the Arm, Shoulder, and Hand Questionnaire revealed a significantly better outcome in the ESWT groups at 3 and 6 months follow-up compared to surgery. Additionally, no adverse events were observed in the ESWT groups, except for small petechiae and haematomas seen in 23 patients, which spontaneously resolved without further treatment. In contrast, complications occurred in 7% (3 patients) of the surgical group. Complications consisted of 2 infections, which needed additional revision surgery and one radial nerve paresis. Based on this level 1 study, the equivalence of ESWT and surgery in stimulating osseous consolidation of long bone nonunions could clearly be demonstrated. Moreover, ESWT seems to be superior in the context of short-term clinical and functional outcomes compared to surgery, entirely avoiding severe complications.

Another study directly comparing ESWT with surgery in nonunions of the 5th metatarsal base performed by Dr. Furia et al. [61] revealed similar results. The ESWT protocol consisted in a single application of 2,000–4,000 impulses at an energy flux density of 0.35 mJ/mm^2 (electrohydraulic, $n = 23$). Surgery in the comparison group ($n = 20$) was performed by closed reduction and intramedullary screw fixation. At final radiological follow-up (6 months), there were no statistical differences in bony union between the two groups (91% union rate in the ESWT group vs. 90% in the surgical group). However, ESWT showed no complications except one case of minor petechiae, whereas surgery overall led to 11 complications. Complications in the surgical group comprised one refracture, one infection as well as nine cases of irritations due to the osteosynthesis material, all requiring additional surgery. Based on this study, we concluded, that both ESWT and screw fixation are suitable in treating fracture nonunions of the fifth metatarsal base. The prevention of complications however depicts a key advantage of ESWT and could prove financially superior to surgery.

The carpal bones, above all the scaphoid, represent another small bone prone to develop nonunions after fracture. Dr. Notarnicola et al. [62] included primarily conservative treated scaphoid fractures, which showed no osseous consolidation on X-rays for a minimum of 6 months after plaster cast immobilization in their retrospective study. These nonunions were treated either with ESWT (n = 58) or those with nonunions underwent surgery (n = 60). ESWT parameters were chosen with 4,000 impulses of electromagnetic source at a mean energy flux density of 0.09 mJ/mm^2. Due to the experiences showing that electromagnetic devices require multiple applications, treatment was scheduled three times at an interval of 72 h. For surgery, the Matti-Russe approach with a cortico-cancellous bone graft was performed. Both groups were treated with cast immobilization after the respective intervention. Again, evaluating bony consolidation as the primary outcome parameter at 12 months after treatment, the authors found comparable results, showing osseous healing of the nonunion in 79% of the cases in the ESWT and in 78% in the surgery group. The Mayo wrist score, as a clinical parameter, comparing ESWT to surgery revealed excellent and good results in 57 and 60% respectively. In contrast to our study, no complications were encountered – neither in the ESWT nor in the surgical group.

Unfortunately due to the relative low incidence of nonunions overall and ethical considerations, RCTs comparing ESWT to an untreated control group were not yet performed. However, the studies mentioned above demonstrated similar to superior results of ESWT in fracture nonunions compared to the standard-of-care treatment being revision surgery to date.

Lacking a control group, further studies investigating the effects of ESWT on fracture nonunions have been carried out. A literature review by Dr. Zelle et al. [63] identified 10 studies (case series of level 4 evidence) including 924 patients of both delayed healings and nonunions. Their review yields an overall healing rate of 76% (95% CI 73–79). Interestingly, comparing data from atrophic and hypertrophic nonunions, only 29% of the atrophic (9 out of 31) and 76% of the hypertrophic (185 out of 243) nonunions achieved bony consolidation.

Since this review, performed in 2010, a number of additional studies evaluating the effect of ESWT for non-healing fractures were performed. Among these, Dr. Alkhawashki [64] treated a total number of 49 nonunions, 38 of them were treated once, 9 twice and 2 three times with 2,000–4,000 impulses depending on the anatomic location at a voltage of 26 kV. ESWT resulted in a union rate of 75.5% after a mean time of 10 months following treatment. Detailed analysis revealed mechanical instability, a fracture gap larger than 5 mm, the type of treated bone (scaphoid), and previously undetected low-grade infection as cause for treatment failure. In contrast, in a study of tibial nonunions, we could show that even infected nonunions responded equally to ESWT compared to uninfected cases (88.5% healing rate) [65]. Similarly, subgroup analysis of hypertrophic vs. oligo/atrophic nonunions, unveiled no statistical differences (94 vs. 88% respectively) considering bony union after ESWT, which somehow contradicts the findings of a study by Dr. Vulpiani et al. [65] and the review conducted by Dr. Zelle et al. [63, 66]. Nonetheless, a number of other studies confirm our findings that nonunion appearance (atrophic/oligotrophic, hypertrophic) is of inferior importance when considering ESWT [67–69]. Our data suggested that ESWT should be performed in a timely manner after the last surgery or trauma in order to increase healing probability [65]. This conclusion was shared by studies performed by Dr. Stojadinovic et al. [67] and Dr. Elster et al. [68].

Looking at tobacco misuse as negative influence on osseous consolidation, as stated in numerous previous reports, Dr. Everding et al. [69] analyzed its impact on healing after ESWT and found no differences in both groups comparing smokers and non-smokers. Dr. Stojadinovicet al. [67] screened for further variables in order to establish a prognostic Naïve Bayesian Classifier influencing outcome after ESWT for nonunions and identified prior intramedullary stabilization the number of previous bone grafting and shockwave treatments as well as anatomical location as relevant predictive factors [67].

According to our own experience with more than 3,500 cases, a clear relationship of the anatomic region/afflicted bone and healing after ESWT could be seen. While tibial nonunions performed best, the scaphoid showed only a union rate of approximately 60% (unpublished data) [65]. Dr. Kuo et al. [70] in their retrospective study demonstrated a healing rate of 64% in atrophic diaphyseal nonunions of the femur ($n = 22$) treated with shockwaves. Besides potential differences in the generation of shockwaves and in energy flux densities, our database as open prospective source yields a healing rate of approximately 75% in that very same location (unpublished data).

Based on the numerous studies investigating the effects of ESWT on delayed or non-healing fractures, along with data from our own collection, which is probably the largest in the world, we conclude that ESWT is a valuable and cost-effective measure, proposing it as first-line therapy for delayed healings or nonunions.

Economic Aspects and Considerations of ESWT in the Treatment of Delayed or Nonunion Fractures

Established nonunions are highly demanding for patients as well as for the treating surgeons. Additionally they imply high socioeconomic burdens to our societies, represented in direct and indirect costs. Directs costs usually are easier to calculate as costs of treatment are somehow easier to compare. Indirect costs, which also reflect costs being spent on rehabilitation as well as on the fact that patients affected by nonunions, at least temporarily lose their ability to work, are in general equally high if not higher than direct costs. Studies on economic aspects in the different countries estimate direct costs arising from the treatment of nonunions ranging from CAD 11,800,

Table 2. Direct cost estimation per year originating from surgery of nonunions in Austria

Number of surgeries	Number of nonunions	Percentage	Costs/case, EUR	Total amount, EUR
1 time	4,760	68	10,000	47,600,000
2 times	1,610	23	20,000	32,200,000
3 times	490	7	30,000	14,700,000
+ Amputation	140	2	40,000	5,600,000
	7,000 cases			100,100,000

2.5% (*n* = 10,000) nonunions of 400,000 fractures annually in Austria, of those 70% being suitable for ESWT (7,000 cases).

USD 11,333 to GBP 29,204 [71–73]. Direct costs in general comprise the expenses related to the treatment of the disease. However, it has to be taken into account that different countries have different settlement modalities, thus making comparison between countries difficult.

In Austrian institutions of the worksmen compensation board (AUVA), an almost constant incidence of nonunions of approximately 2.5% of all fractures (around 10,000 nonunions of 400,000 fractures/year) can be noted over the years. In order to estimate costs related to fracture-healing failures in Austria, the AUVA conducted a cost approximation supported by the University of Economics in Vienna (Viennese business school, WU Wien). Based on general assumptions, healing was hypothesized in 68% of nonunions after the first surgical intervention and in 23% after a second surgical intervention. Of the remaining 9%, 7% were presumed to finally heal after a third intervention and 2% were considered to end up in amputation. Around 70% of the nonunions occurring annually in Austria are suitable for ESWT (about 7,000 patients). Based on recent literature and on our own experience, 75% of the nonunions could be successfully treated with ESWT. Reflecting our own experience and based on statistical evaluation of our database over the years, around 25% of these will potentially need a second ESWT if using an electrohydraulic device for treatment. Mean costs for surgical interventions for these 7,000 nonunion cases per year, which could potentially be treated with ESWT, would amount to EUR 100,100,000. These costs are only an estimation and are based on the assumption that surgical intervention would approximately cost EUR 10,000/case (Table 2). Comparing these costs to the costs mentioned in literature, our presumption tends to be underestimated. However, these costs are only taking direct costs into consideration. ESWT on the other hand, if calculated with EUR 1,000/treated nonunion case would sum up to a total cost of EUR 33,780,000. In this calculation, the surgical costs for nonunions recalcitrant to ESWT even after a second session, are included, and were estimated as stated above with 25% which yields 1,750 cases (Table 3). Subtracting costs of ESWT of the costs of surgical interventions, would lead to an overall reduction of EUR 66,320,000 in direct costs.

Table 3. Direct cost estimation per year originating from ESWT (+surgery in recalcitrant cases) of nonunions in Austria

	Number of nonunions	Percentage	Costs/case, EUR	Total amount, EUR
Number of ESWT sessions				
1 time	5,250	75	1,000	5,250,000
2 times	1,750	25	1,000	1,750,000
Recalcitrant cases	1,750	25	1,000	1,750,000
	7,000 cases			8,750,000
Number of surgeries				
1 time	1,190	68	10,000	11,900.000
2 times	402	23	20,000	8,040,000
3 times	123	7	30,000	3,690,000
+ Amputation	35	2	40,000	1,400,000
	1,750 cases			25,030,000

1,750 cases not responding to ESWT are further treated surgically. 2.5% (n = 10,000) nonunions of 400,000 fractures annually in Austria, of those 70% being suitable for ESWT (7,000 cases); 25% (= 1,750 cases) treated with ESWT are nonresponders, thus subjected to surgery.

Hence, ESWT would dramatically reduce costs, saving about 66% of the expenditures spent on treating nonunions/year. These calculations, however, focus just on direct costs, omitting any indirect costs.

The indirect costs, based on a simplified view as being roughly the loss of work force, amounts to approximately 82.8–93.3% of all costs for patients with tibial fractures in the European health care system [74]. Taking this into consideration and based on the fact that the majority of studies showed equal, although faster healing with ESWT compared to surgery, would lead to the fact, that patients could be sooner reintegrated in their working environment, cutting down indirect costs. All together this underlines that ESWT is not only an effective tool in the treatment of nonunions but also an extremely cost-efficient method being superior compared to surgery as standard of care to date.

ESWT as a Potential Treatment Option in Fragility Fractures

The increasing life expectancy consistently leads to an increase in degenerative diseases in the ageing population. Among these, osteoporosis implies high socioeconomic costs to our societies, as fracture rates increase and rehabilitation proves to be lengthier and more costly when compared to the rates of acute fracture care [75]. Osteoporosis, defined as reduction in bone mineral density and deterioration of the bone microarchitecture confirmed by bone histomorphometry, namely, in the decrease of total bone volume, trabecular numbers and trabecular thickness as well as the increase in trabecular separation, finally increases bone fragility. Consequently, fragility fractures increase with age. Above all, in women, a constant increase in fractures can be seen after the age of 60. This postmenopausal increase can be mainly attributed to osteoporosis. Osteoporotic fractures frequently have a devastating impact on the patients' quality of life [76–78]. Mortality rates in these patients are generally increased, which is mostly related to comorbidities and complications of the operative procedures, such as failure of the instrumentation and/or healing disturbances depicted as delayed healings or nonunions, rather than to the fracture itself [76]. Furthermore, the diminished bone stock most often leads to a prolonged rest or inactivation, potentially causing pneumonia or venous thromboembolism. In order to reduce instrumentation failures, locking plates or fixed-angle devices have gained more attention in the area of fixation of osteoporotic fractures. The locking of the screw onto the plate provides angular and axial stability of the construct, even in cases of diminished bone quality. The cold shut of these locking screws with the plate almost entirely avoids sliding or secondary displacement, thus reducing the risk of failure of the osteosynthesis due to a secondary loss of reduction. However, to date, there is no perfect solution for osteoporotic fractures and complications remain to be high, even with the use of the above-mentioned devices. Consequently, research still seeks for methods in order to optimize treatment and outcomes of these fractures. Pharmacological studies have proven that bisphosphonates, RANKL inhibitors, as well as potentially sclerostin antibodies are effective in reducing additional fractures, namely, vertebral and femoral neck fractures in osteoporosis. Nevertheless, literature covering adjunctive methods in order to reduce fixation failure or nonunions is scarce, and as in acute fractures, the use of these measures is always debated upon in order to prevent or reduce complications especially in this high-risk patient group.

Dr. van der Jagt et al. [79] first investigated the influence of ESWT in osteoporotic ovarectomized rodents, looking at dynamic changes of bone microarchitecture following a fibula osteotomy. A single application of 2,000 shockwave impulses at 0.16 mJ/mm^2 yielded significantly higher trabecular bone volume fractions of the proximal tibia compared to the untreated contralateral side, lasting up to 7 weeks after treatment. However, ESWT did not influence the healing of the fibula osteotomy [79]. Another study performed in ovarectomized rats with a proximal tibial osteotomy fixed intramedullarily, however, demonstrated a higher bone volume, bone volume/tissue volume, and trabecular thickness at the osteotomy level compared to untreated controls [80]. SPECT analysis on healthy tibia revealed an increased uptake of technetium-labeled methylene diphosponate in response to a single application of ESWT, suggesting enhanced metabolic activity of osteoblasts. Additionally, structural analysis performed by micro-computertomography (CT) imaging illustrated both increased trabecular and cortical volume, leading to better biomechanical properties in osteoporotic bones [81, 82].

Dr. Koolen et al. [83] used ESWT as an adjunct to improve osseointegration of cortical and cancellous screws, not in osteoporotic but healthy rodent bones. While the cortical screws improved in biomechanical testing, hence better osseointegration was assumed, the cancellous screws showed no differences in testing after ESWT.

While treatment of osteoporosis prophylactically by ESWT might neither be feasible nor financially reasonable, concomitant treatment of fragility fractures with ESWT at the time of surgery could potentially improve and accelerate healing.

Radial Pressure Waves in Bone Indications

While radial pressure wave therapy certainly has its value in the treatment of certain orthopedic diseases, such as superficial tendinopathies and other orthopedic problems in soft tissues, the application in nonunions has to be critically scrutinized. Even though, there are sporadic successful reports on nonunions treated with radial pressure waves, the achieved healing rate is far below than the healing rate of high energy shockwave treatment (whether electrohydraulic or electromagnetic). Subgroup analysis (of our group), within the open prospective trial in the AUVA trauma center Meidling, evaluated the effect of an electrohydraulic shockwave device of lower energy flux densities (OW180C) compared to a device of the same company with a higher energy flux density. The healing rate of the latter compared favorably to the device with the lower energy flux density, yielding in osseous healing in 81 and 66% of the cases respectively. Taking the physical characteristics and the fact that energy decreases with the square of distance into account, it remains highly questionable if bone tissue is sufficiently stimulated by radial pressure waves.

This is supported by an experimental in vitro study investigating radial pressure waves on the activity of osteoblasts [84]. The study showed an inhibited osteoblastogenesis with statistically significant reduction in type 1 collagen, osterix, bone sialoprotein, and receptor activator NF kappa ligand expression. The authors therefore concluded that radial pressure wave therapy is not indicated in the treatment of delayed healings or nonunions of bone. Conversely, Dr. Gollwitzer et al. [85] performed a study in healthy rabbits, applying radial pressure waves twice in an interval of 1 week at 4 bar (corresponding to 0.16 mJ/mm^2) on the ipsilateral femur. The contralateral femur of the rabbits served as internal control. At 4 and 6 weeks after treatment, a higher extent of bone formation was found in a semiquantitative analysis using fluorescence microscopy. Alizarin red and calcein blue bands were counted as positive markers for new bone formation. However, clinical translations of these findings are difficult to apply, as results depict a semi-quantitative analysis of bone formation in healthy animals with only limited soft-tissue envelopes.

Based on the limited data available and on our own experience, we strongly discourage the use of radial pressure wave therapy in treating healing problems in bone so far. Especially from an ethical point of view, patients affected by nonunion should receive the most adequate and a successful therapy available; even then radial pressure wave has a certain effect on bones.

Nonetheless, to finally assess the clinical relevance of radial pressure waves in the indication of nonunions, further RCTs are encouraged and needed.

ESWT in Stress Fractures

Stress fractures (SF) were first described by Breithaupt in Prussian soldiers complaining of leg pain in the war of 1850 [86]. Worldwide incidences of SFs range from 1 to 20% depending on patients' physical activities. They represent about 10% of all injuries encountered in sports medicine [87].

Clinically they present as progressive localized pain after physical activity or sports, which usually quickly resolves with rest. SFs represent a classic overuse or repetitive strain injury and hence are frequently encountered in athletes and military personnel. They are best described as continuum pathology, beginning with bone marrow edema followed by trabecular fracture and cortical disruption peaking in complete fracture of the affected bone. These fractures usually occur as soon as unphysiological mechanical loads of the affected bone exceed its intrinsic loading capacity. This excessive exposure to mechanical loads occurs over time. In contrast, so-called insufficiency/ fragility/osteoporotic or pathological fractures occur if normal loads lead to fracture. In these cases, an underlying pathology results in a diminished bone quality making the bone prone to fracture.

According to the American College of Radiology, MRI is considered gold standard in imaging of SFs. Above all, in early stages where X-rays are typically negative, MRI is equally sensitive (100%) to bone scintigraphy with higher specificity (85%) in detecting SFs [87].

SFs can be classified based on their risk for healing failures into low, medium, and high risk (Table 4). High-risk SF usually needs surgical management to prevent complete fracture, whereas low-risk SF usually responds to conservative treatment. Conservative measures include rest, ice, and pain control, the latter being usually realized with non-steroidal anti-inflammatory drugs in order to control pain. Nevertheless, these drugs should be used with caution, as they have the potential to negatively interfere with bone healing. Rest is frequently debated as it has never been properly investigated. As in repetitive strain injuries of soft tissues as delineated in the continuum theory of the Dr. Cook and Purdam [88], complete rest could potentially be harmful leading to stress shielding. Thereafter in recent days, protection and load modifications as best described in optimal loading should be advocated.

However, about 1/3 of low risk SFs will eventually continue to be painful during exercise after conservative treatment and hence progress to a higher risk class. Especially in these cases, we might advocate ESWT as treatment of first choice.

　　　　　　　　　　　　　　　　　　　Haffner·Smolen·Dahm·Schaden·Mittermayr

Table 4. Grading of anatomic locations at risk of stress fractures

Low risk	Moderate risk	High risk
2–4th metatarsal shaft	Pelvis	Pars interarticularis of the lumbar spine
Fibula and lateral malleolus	Femoral shaft	Femoral head
Cuneiforms	Postero-medial tibia	Lateral femoral neck
Calcaneus	Medial malleolus	Patella
Cuboid	Base of 5th metatarsal	Anterior shape of the tibia
Medial femoral neck		Proximal 2nd metatarsal Great toe Hallux sesamoids Tarsal bones

Treatment of SFs using ESWT is a new approach lacking sufficient studies yet. After an anecdotal report of the Dr. Hotzinger [89] at the ISMST meeting in London in 1999, applying high-energy shockwave treatment in tibial SFs, this entity came into the focus of interest. Dr. Leal [90] conducted an experimental randomized, single-blinded, self-controlled clinical trial on 26 eighteen-year-old navy cadets with identical bilateral tibial SFs in 2001. SFs were symptomatic for more than 3 months and responded poorly to conventional treatment. In the treatment group, 2,000 impulses were applied in 2 sessions 1 week apart. The energy flux density ranged between 0.1 and 0.27 mJ/mm^2. They concluded that ESWT significantly reduced pain and recovery time in naval cadets with tibial SFs. The same group reported a good outcome in treating an SF of the navicular bone of a gymnast.

Similarly, Dr. Audain [91] and Dr. Gordon [92] reported on several cases of SFs in high-performing athletes, yielding good results with respect to pain control and return to competition. In 2007, Dr. Taki et al. [93] studied SFs unresponsive to conventional therapy in 5 athletes. In contrast to the other studies, he used considerably higher energy flux densities of 0.29–0.40 mJ/mm^2, applied once with 2,000–40,000 impulses. By doing so, he was able to significantly reduce the recovery time to 3–6 months. Dr. Moretti et al. [94] in a study conducted in 2009 reported their results of treating 10 high performance athletes with SFs of tibias or metatarsals. As in their case report series with acute fractures, their application was quite different to all of the above-mentioned studies. Their protocol used 4,000 shockwaves of 0.09–0.17 mJ/mm^2 each applied in 3–4 sessions. In the 8 weeks post treatment, a healing rate of 100% was observed and all patients returned to prior competition levels.

Since 1/3 of low-risk SF will eventually progress to a higher risk SF, ESWT can be recommended in low-risk SF if no improvement is seen after conservative treatment. Moderate-risk SF should be included in this algorithm and ESWT might be advocated even earlier, as progression to high-risk SF after failure to respond to conservative

Table 5. Recommendations about contraindications using high-energy focused shockwaves by the ISMST (International Society for Medical Shockwave Treatment; www.shockwavetherapy.org)

Lung tissue in the treatment area
Malignant tumor in the treatment area (not as underlying disease)
Epiphyseal plate in the treatment area
Brain or Spine in the treatment area
Severe coagulopathy
Fetus in the treatment area

Close attention should be paid to the depth of penetration of shockwaves when treating deep tissue structures.

treatment remains possible. High-risk SF, which to date is mostly operated, could potentially benefit from ESWT, avoiding complications associated with surgery. However, in order to elucidate the role of ESWT in the treatment of SFs further RCTs are needed.

Evaluation and Clinical Performance of ESWT in Bone Pathologies

In order to ensure adequate prerequisites for ESWT, it is essential to accurately evaluate the underlying bone pathology prior to treatment. The suitability of the nonunion intended to treat has to be carefully evaluated concerning the indications and contraindications published in an updated version by the ISMST (Table 5). Typical symptoms of patients suffering from nonunions include local heat, swelling, redness, pain to compression, and bending as well as abnormal movements of bone fragments. Imaging studies should comprise at least conventional radiographs in two planes. Additional imaging such as a CT scan can be helpful to quantify the extent of the delayed or non-healing fracture. Furthermore, the degree of the osseous consolidation prior to ESWT can be assessed. Three-dimensional reconstructions can be of particular help not only prior to treatment but also during further radiological follow-up. In cases of SFs, bone marrow edema, or fatigue fractures, an MRI study might be of superior relevance. This also accounts true for fracture fragments, which are of questionable vitality. Clinical data suggest that a bone fragment diastasis (in long bones) of more than 0.5 cm lead to reduced efficacy of ESWT. ESWT applied to bones preferentially imply focused shockwaves at high-energy flux densities. High-energy flux densities would cause pain, hence ESWT in this setting, in contrast to application of radial or ballistic waves in soft tissues, requires general or regional anesthesia. Local anesthetics, should be avoided, as it was shown to reduce efficacy of ESWT. In our clinic/institution, general anesthesia using a laryngeal mask is the preferred mode of narcosis, as it is well tolerated and controlled in an easy manner. Approximately, 500 impulses prior to the termination of ESWT, anesthesia is diverted, normally allowing the patient to be awake by the end of the therapy. As ESWT itself has some analgesic effects, postinterventional analgesia is rarely needed.

The area of interest (i.e., the fracture gap) is located by X-ray and the fracture and its configuration is marked on the skin by a permanent marker, allowing adjustments of the therapy head during the treatment without losing focus on the fracture site. Thereafter, an adequate amount of bubble-free conduction gel is applied on the area of interest and the therapy head is accurately positioned. Subsequently, coupling of the therapy head at the fracture site is achieved and the intended amounts of impulses are applied. In our clinic, usually 2,000–4,000 impulses at an energy level of 0.3–0.4 mJ/mm^2 at 1–5 Hertz are applied (electrohydraulic device). The fracture gap should be targeted from at least 3 different directions. According to our own studies, osteosynthetic material at the fracture site did not interfere either with the application of ESWT or with the outcome. A completion of a detailed OR protocol is recommended in order to better compare outcomes and to delineate deviations from the currently known (side) effects of ESWT. Occasionally, reddening and petechial bleeding were observed immediately after therapy, which resolved within a couple of days, without further interventions. When using an electrohydraulic device, only one treatment is scheduled. In contrast, ESWT applied with an electromagnetic device usually requires 2–4 treatment in a weekly or biweekly interval (0.4–0.6 mJ/mm^2, 4,000–6,000 impulses).

Aftercare follows the principles of acute fractures treated conservatively. Stability in general is realized by a plaster cast, orthosis, or splint. Weight-bearing is usually reduced to partial or non-weight bearing in the first 4 weeks after ESWT as deemed appropriate by the treating physician/surgeon. Limitation of weight bearing does not necessarily rely on fracture stability but more on the fact that neovascularization or neoangiogenesis might be disturbed at the fracture site. Subsequent X-rays are anticipated every month, until bridging of at least 3 of 4 cortices is achieved. In our institution, CT scans are performed after 3 and 6 months following ESWT in order to judge healing in a more detailed manner. To ensure compliance of the patient, it is essential to inform patients that clinical and radiological follow-up after ESWT will last up to 6 months, to finally evaluate osseous healing. If nonunion is still present after 6 months, reevaluation of the pathology and possible causes for failure is undertaken and further options are discussed with the patient. Further options include a secondary ESWT or revision surgery. As mentioned above, about 1/4 of the cases will heal after a second ESWT.

Concluding Remarks

Overall, ESWT has been shown to be a reliable, safe, and highly effective treatment especially in delayed or non-healing fractures. Due to the fact that it entirely avoids severe complications as seen in surgical cases, at least in Austria it has been imposed as first-line treatment in these cases. Furthermore, it represents a treatment option in SFs at higher risk, not responding to conservative measures. Its adjunctive value for acute as well as for fragility fractures has to be seen in the future, while we wistfully await further RCTs to finally clarify the role of ESWT in these indications.

References

1 Mathew G, Hanson BP: Global burden of trauma: need for effective fracture therapies. Indian J Orthop 2009;43:111–116.
2 Sprague S, Bhandari M: An economic evaluation of early versus delayed operative treatment in patients with closed tibial shaft fractures. Arch Orthop Trauma Surg 2002;122:315–323.
3 Taylor JC: Delayed union and nonunion of fractures; in Crenshaw AH (ed): Campbell's Operative Orthopaedics, ed 8. St. Louis, MO, Mosby, 1992, pp 1287–1345.
4 Hak DJ: Management of aseptic tibial nonunion. J Am Acad Orthop Surg 2011;19:563–573.
5 Becker ST, Warnke PH, Behrens E, Wiltfang J: Morbidity after iliac crest bone graft harvesting over an anterior versus posterior approach. J Oral Maxillofac Surg 2011;69:48–53.
6 Nolte PA, van der Krans A, Patka P, Janssen IM, Ryaby JP, Albers GH: Low-intensity pulsed ultrasound in the treatment of nonunions. J Trauma 2001;51:693–702; discussion 702–693.
7 Gebauer D, Mayr E, Orthner E, Ryaby JP: Low-intensity pulsed ultrasound: effects on nonunions. Ultrasound Med Biol 2005;31:1391–1402.
8 Heckman JD, Ryaby JP, McCabe J, Frey JJ, Kilcoyne RF: Acceleration of tibial fracture-healing by non-invasive, low-intensity pulsed ultrasound. J Bone Joint Surg Am 1994;76:26–34.
9 Nelson FR, Brighton CT, Ryaby J, Simon BJ, Nielson JH, Lorich DG, Bolander M, Seelig J: Use of physical forces in bone healing. J Am Acad Orthop Surg 2003;11:344–354.
10 Heckman JD, Ingram AJ, Loyd RD, Luck JV Jr, Mayer PW: Nonunion treatment with pulsed electromagnetic fields. Clin Orthop Relat Res 1981;161:58–66.
11 Garland DE, Moses B, Salyer W: Long-term follow-up of fracture nonunions treated with PEMFs. Contemp Orthop 1991;22:295–302.
12 Kanakaris NK, Calori GM, Verdonk R, Burssens P, De Biase P, Capanna R, Vangosa LB, Cherubino P, Baldo F, Ristiniemi J, Kontakis G, Giannoudis PV: Application of BMP-7 to tibial non-unions: a 3-year multicenter experience. Injury 2008;39(suppl 2):S83–S90.
13 Kanakaris NK, Paliobeis C, Nlanidakis N, Giannoudis PV: Biological enhancement of tibial diaphyseal aseptic non-unions: the efficacy of autologous bone grafting, bmps and reaming by-products. Injury 2007;38(suppl 2):S65–S75.
14 Pecina M, Haspl M, Jelic M, Vukicevic S: Repair of a resistant tibial non-union with a recombinant bone morphogenetic protein-7 (rh-BMP-7). Int Orthop 2003;27:320–321.
15 Haupt G, Haupt A, Gerety B, Chevapil M: Enhancement of fracture healing with extracopreal shock wave. J Urol 1990;143:23.
16 Valchanou VD, Michailov P: High energy shock waves in the treatment of delayed and nonunion of fractures. Int Orthop 1991;15:181–184.
17 Tischer T, Milz S, Weiler C, Pautke C, Hausdorf J, Schmitz C, Maier M: Dose-dependent new bone formation by extracorporeal shock wave application on the intact femur of rabbits. Eur Surg Res 2008;41:44–53.
18 Schaden W, Fischer A, Sailler A: Extracorporeal shock wave therapy of nonunion or delayed osseous union. Clin Orthop Relat Res 2001;387:90–94.
19 Wang CJ, Wang FS, Yang KD, Weng LH, Hsu CC, Huang CS, Yang LC: Shock wave therapy induces neovascularization at the tendon-bone junction. A study in rabbits. J Orthop Res 2003;21:984–989.
20 Wang CJ, Wang FS, Yang KD: Biological effects of extracorporeal shockwave in bone healing: a study in rabbits. Arch Orthop Trauma Surg 2008;128:879–884.
21 Huang C, Holfeld J, Schaden W, Orgill D, Ogawa R: Mechanotherapy: revisiting physical therapy and recruiting mechanobiology for a new era in medicine. Trends Mol Med 2013;19:555–564.
22 Ha CH, Kim S, Chung J, An SH, Kwon K: Extracorporeal shock wave stimulates expression of the angiogenic genes via mechanosensory complex in endothelial cells: mimetic effect of fluid shear stress in endothelial cells. Int J Cardiol 2013;168:4168–4177.
23 Xu JK, Chen HJ, Li XD, Huang ZL, Xu H, Yang HL, Hu J: Optimal intensity shock wave promotes the adhesion and migration of rat osteoblasts via integrin β1-mediated expression of phosphorylated focal adhesion kinase. J Biol Chem 2012;287:26200–26212.

24 Ingber DE: Cellular mechanotransduction: putting all the pieces together again. FASEB J 2006;20:811–827.

25 Hu J, Liao H, Ma Z, Chen H, Huang Z, Zhang Y, Yu M, Chen Y, Xu J: Focal adhesion kinase signaling mediated the enhancement of osteogenesis of human mesenchymal stem cells induced by extracorporeal shockwave. Sci Rep 2016;6:20875.

26 Martini L, Giavaresi G, Fini M, Torricelli P, Borsari V, Giardino R, De Pretto M, Remondini D, Castellani GC: Shock wave therapy as an innovative technology in skeletal disorders: Study on transmembrane current in stimulated osteoblast-like cells. Int J Artif Organs 2005;28:841–847.

27 Chen YJ, Kuo YR, Yang KD, Wang CJ, Sheen Chen SM, Huang HC, Yang YJ, Yi-Chih S, Wang FS: Activation of extracellular signal-regulated kinase (ERK) and p38 kinase in shock wave-promoted bone formation of segmental defect in rats. Bone 2004;34:466–477.

28 Wang FS, Wang CJ, Chen YJ, Chang PR, Huang YT, Sun YC, Huang HC, Yang YJ, Yang KD: Ras induction of superoxide activates erk-dependent angiogenic transcription factor HIF-1alpha and VEGF-A expression in shock wave-stimulated osteoblasts. J Biol Chem 2004;279:10331–10337.

29 Wang FS, Wang CJ, Huang HJ, Chung H, Chen RF, Yang KD: Physical shock wave mediates membrane hyperpolarization and ras activation for osteogenesis in human bone marrow stromal cells. Biochem Biophys Res Commun 2001;287:648–655.

30 Wang FS, Wang CJ, Sheen-Chen SM, Kuo YR, Chen RF, Yang KD: Superoxide mediates shock wave induction of ERK-dependent osteogenic transcription factor (CBFA1) and mesenchymal cell differentiation toward osteoprogenitors. J Biol Chem 2002;277:10931–10937.

31 Tamma R, dell'Endice S, Notarnicola A, Moretti L, Patella S, Patella V, Zallone A, Moretti B: Extracorporeal shock waves stimulate osteoblast activities. Ultrasound Med Biol 2009;35:2093–2100.

32 Hofmann A, Ritz U, Hessmann MH, Alini M, Rommens PM, Rompe JD: Extracorporeal shock wave-mediated changes in proliferation, differentiation, and gene expression of human osteoblasts. J Trauma 2008;65:1402–1410.

33 Takahashi K, Yamazaki M, Saisu T, Nakajima A, Shimizu S, Mitsuhashi S, Moriya H: Gene expression for extracellular matrix proteins in shockwave-induced osteogenesis in rats. Calcif Tissue Int 2004;74:187–193.

34 Tam K-F, Cheung W-H, Lee K-M, Qin L, Leung K-S: Osteogenic effects of low-intensity pulsed ultrasound, extracorporeal shockwaves and their combination – an in vitro comparative study on human periosteal cells. Ultrasound Med Biol 2008;34:1957–1965.

35 Tam KF, Cheung WH, Lee KM, Qin L, Leung KS: Delayed stimulatory effect of low-intensity shockwaves on human periosteal cells. Clin Orthop Relat Res 2005;438:260–265.

36 Dias dos Santos PR, De Medeiros VP, Freire Martins de Moura JP, da Silveira Franciozi CE, Nader HB, Faloppa F: Effects of shock wave therapy on glycosaminoglycan expression during bone healing. Int J Surg 2015;24:120–123.

37 Huang HM, Li XL, Tu SQ, Chen XF, Lu CC, Jiang LH: Effects of roughly focused extracorporeal shock waves therapy on the expressions of bone morphogenetic protein-2 and osteoprotegerin in osteoporotic fracture in rats. Chin Med J (Engl) 2016;129:2567–2575.

38 Wang FS, Yang KD, Kuo YR, Wang CJ, Sheen-Chen SM, Huang HC, Chen YJ: Temporal and spatial expression of bone morphogenetic proteins in extracorporeal shock wave-promoted healing of segmental defect. Bone 2003;32:387–396.

39 Chen YJ, Kuo YR, Yang KD, Wang CJ, Huang HC, Wang FS: Shock wave application enhances pertussis toxin protein-sensitive bone formation of segmental femoral defect in rats. J Bone Miner Res 2003;18:2169–2179.

40 Wang FS, Yang KD, Chen RF, Wang CJ, Sheen-Chen SM: Extracorporeal shock wave promotes growth and differentiation of bone-marrow stromal cells towards osteoprogenitors associated with induction of TGF-beta1. J Bone Joint Surg Br 2002;84:457–461.

41 Iannone F, Moretti B, Notarnicola A, Moretti L, Patella S, Patella V, Lapadula G: Extracorporeal shock waves increase interleukin-10 expression by human osteoarthritic and healthy osteoblasts in vitro. Clin Exp Rheumatol 2009;27:794–799.

42 Sun D, Junger WG, Yuan C, Zhang W, Bao Y, Qin D, Wang C, Tan L, Qi B, Zhu D, Zhang X, Yu T: Shockwaves induce osteogenic differentiation of human mesenchymal stem cells through atp release and activation of P2X7 receptors. Stem Cells 2013;31:1170–1180.

43 Weihs AM, Fuchs C, Teuschl AH, Hartinger J, Slezak P, Mittermayr R, Redl H, Junger WG, Sitte HH, Runzler D: Shock wave treatment enhances cell proliferation and improves wound healing by atp release-coupled extracellular signal-regulated kinase (ERK) activation. J Biol Chem 2014;289:27090–27104.

44 Xu L, Zhao Y, Wang M, Song W, Li B, Liu W, Jin X, Zhang H: Defocused low-energy shock wave activates adipose tissue-derived stem cells in vitro via multiple signaling pathways. Cytotherapy 2016;18:1503–1514.

45 Hatanaka K, Ito K, Shindo T, Kagaya Y, Ogata T, Eguchi K, Kurosawa R, Shimokawa H: Molecular mechanisms of the angiogenic effects of low-energy shock wave therapy: Roles of mechanotransduction. Am J Physiol Cell Physiol 2016;311:C378–C385.

46 Yu W, Shen T, Liu B, Wang S, Li J, Dai D, Cai J, He Q: Cardiac shock wave therapy attenuates H9C2 myoblast apoptosis by activating the akt signal pathway. Cell Physiol Biochem 2014;33:1293–1303.

47 Sheu JJ, Lee FY, Yuen CM, Chen YL, Huang TH, Chua S, Chen YL, Chen CH, Chai HT, Sung PH, Chang HW, Sun CK, Yip HK: Combined therapy with shock wave and autologous bone marrow-derived mesenchymal stem cells alleviates left ventricular dysfunction and remodeling through inhibiting inflammatory stimuli, oxidative stress and enhancing angiogenesis in a swine myocardial infarction model. Int J Cardiol 2015;193:69–83.

48 Tepekoylu C, Wang FS, Kozaryn R, Albrecht-Schgoer K, Theurl M, Schaden W, Ke HJ, Yang Y, Kirchmair R, Grimm M, Wang CJ, Holfeld J: Shock wave treatment induces angiogenesis and mobilizes endogenous cd31/cd34-positive endothelial cells in a hindlimb ischemia model: Implications for angiogenesis and vasculogenesis. J Thorac Cardiovasc Surg 2013;146:971–978.

49 Fu M, Sun CK, Lin YC, Wang CJ, Wu CJ, Ko SF, Chua S, Sheu JJ, Chiang CH, Shao PL, Leu S, Yip HK: Extracorporeal shock wave therapy reverses ischemia-related left ventricular dysfunction and remodeling: Molecular-cellular and functional assessment. PLoS One 2011;6:e24342.

50 Schuh CM, Heher P, Weihs AM, Banerjee A, Fuchs C, Gabriel C, Wolbank S, Mittermayr R, Redl H, Runzler D, Teuschl AH: In vitro extracorporeal shock wave treatment enhances stemness and preserves multipotency of rat and human adipose-derived stem cells. Cytotherapy 2014;16:1666–1678.

51 Zhai L, Ma XL, Jiang C, Zhang B, Liu ST, Xing GY: Human autologous mesenchymal stem cells with extracorporeal shock wave therapy for nonunion of long bones. Indian J Orthop 2016;50:543–550.

52 Kearney CJ, Hsu HP, Spector M: The use of extracorporeal shock wave-stimulated periosteal cells for orthotopic bone generation. Tissue Eng Part A 2012;18:1500–1508.

53 Wang CJ, Liu HC, Fu TH: The effects of extracorporeal shockwave on acute high-energy long bone fractures of the lower extremity. Arch Orthop Trauma Surg 2007;127:137–142.

54 Moretti B, Notarnicola A, Moretti L, Patella S, Tato I, Patella V: Bone healing induced by ESWT. Clin Cases Miner Bone Metab 2009;6:155–158.

55 Kieves NR, MacKay CS, Adducci K, Rao S, Goh C, Palmer RH, Duerr FM: High energy focused shock wave therapy accelerates bone healing. A blinded, prospective, randomized canine clinical trial. Vet Comp Orthop Traumatol 2015;28:425–432.

56 Beutler S, Regel G, Pape HC, Machtens S, Weinberg AM, Kremeike I, Jonas U, Tscherne H: [extracorporeal shock wave therapy for delayed union of long bone fractures – preliminary results of a prospective cohort study]. Unfallchirurg 1999;102:839–847.

57 Ikeda K, Tomita K, Takayama K: Application of extracorporeal shock wave on bone: Preliminary report. J Trauma 1999;47:946–950.

58 Rompe JD, Rosendahl T, Schollner C, Theis C: High-energy extracorporeal shock wave treatment of nonunions. Clin Orthop Relat Res 2001;387:102–111.

59 Wang CJ, Chen HS, Chen CE, Yang KD: Treatment of nonunions of long bone fractures with shock waves. Clin Orthop Relat Res 2001;387:95–101.

60 Cacchio A, Giordano L, Colafarina O, Rompe JD, Tavernese E, Ioppolo F, Flamini S, Spacca G, Santilli V: Extracorporeal shock-wave therapy compared with surgery for hypertrophic long-bone nonunions. J Bone Joint Surg Am 2009;91:2589–2597.

61 Furia JP, Juliano PJ, Wade AM, Schaden W, Mittermayr R: Shock wave therapy compared with intramedullary screw fixation for nonunion of proximal fifth metatarsal metaphyseal-diaphyseal fractures. J Bone Joint Surg Am 2010;92:846–854.

62 Notarnicola A, Moretti L, Tafuri S, Gigliotti S, Russo S, Musci L, Moretti B: Extracorporeal shockwaves versus surgery in the treatment of pseudoarthrosis of the carpal scaphoid. Ultrasound Med Biol 2010;36:1306–1313.

63 Zelle BA, Gollwitzer H, Zlowodzki M, Buhren V: Extracorporeal shock wave therapy: Current evidence. J Orthop Trauma 2010;24(suppl 1):S66–S70.

64 Alkhawashki HM: Shock wave therapy of fracture nonunion. Injury 2015;46:2248–2252.

65 Haffner N, Antonic V, Smolen D, Slezak P, Schaden W, Mittermayr R, Stojadinovic A: Extracorporeal shockwave therapy (ESWT) ameliorates healing of tibial fracture non-union unresponsive to conventional therapy. Injury 2016;47:1506–1513.

66 Vulpiani MC, Vetrano M, Conforti F, Minutolo L, Trischitta D, Furia JP, Ferretti A: Effects of extracorporeal shock wave therapy on fracture nonunions. Am J Orthop (Belle Mead NJ) 2012;41:E122–E127.

67 Stojadinovic A, Kyle Potter B, Eberhardt J, Shawen SB, Andersen RC, Forsberg JA, Shwery C, Ester EA, Schaden W: Development of a prognostic naive bayesian classifier for successful treatment of nonunions. J Bone Joint Surg Am 2011;93:187–194.

68 Elster EA, Stojadinovic A, Forsberg J, Shawen S, Andersen RC, Schaden W: Extracorporeal shock wave therapy for nonunion of the tibia. J Orthop Trauma 2010;24:133–141.

69 Everding J, Freistuhler M, Stolberg-Stolberg J, Raschke MJ, Garcia P: [extracorporal shock wave therapy for the treatment of pseudarthrosis: new experiences with an old technology]. Unfallchirurg 2016;120:969–978.

70 Kuo SJ, Su IC, Wang CJ, Ko JY: Extracorporeal shockwave therapy (ESWT) in the treatment of atrophic non-unions of femoral shaft fractures. Int J Surg 2015;24:131–134.

71 Beaver R, Brinker MR, Barrack RL: An analysis of the actual cost of tibial nonunions. J La State Med Soc 1997;149:200–206.

72 Busse JW, Bhandari M, Sprague S, Johnson-Masotti AP, Gafni A: An economic analysis of management strategies for closed and open grade I tibial shaft fractures. Acta Orthopaedica 2009;76:705–712.

73 Patil S: Management of complex tibial and femoral nonunion using the ilizarov technique, and its cost implications. J Bone Joint Surg Br 2006;88:928–932.

74 Hak DJ, Fitzpatrick D, Bishop JA, Marsh JL, Tilp S, Schnettler R, Simpson H, Alt V: Delayed union and nonunions: epidemiology, clinical issues, and financial aspects. Injury 2014;45:S3–S7.

75 Curtis EM, Moon RJ, Harvey NC, Cooper C: The impact of fragility fracture and approaches to osteoporosis risk assessment worldwide. Bone 2017;104:29–38.

76 Sambrook P, Cooper C: Osteoporosis. Lancet 2006;367:2010–2018.

77 Bliuc D: Mortality risk associated with low-trauma osteoporotic fracture and subsequent fracture in men and women. JAMA 2009;301:513.

78 Johnell O, Kanis JA: An estimate of the worldwide prevalence and disability associated with osteoporotic fractures. Osteoporos Int 2006;17:1726–1733.

79 van der Jagt OP, van der Linden JC, Schaden W, van Schie HT, Piscaer TM, Verhaar JAN, Weinans H, Waarsing JH: Unfocused extracorporeal shock wave therapy as potential treatment for osteoporosis. J Orthop Res 2009;27:1528–1533.

80 Chen XF, Huang HM, Li XL, Liu GJ, Zhang H: Slightly focused high-energy shockwave therapy: a potential adjuvant treatment for osteoporotic fracture. Int J Clin Exp Med 2015;8:5044–5054.

81 van der Jagt OP, Piscaer TM, Schaden W, Li J, Kops N, Jahr H, van der Linden JC, Waarsing JH, Verhaar JN, de Jong M, Weinans H: Unfocused extracorporeal shock waves induce anabolic effects in rat bone. J Bone Joint Surg Am 2011;93:38–48.

82 van der Jagt OP, Waarsing JH, Kops N, Schaden W, Jahr H, Verhaar JA, Weinans H: Unfocused extracorporeal shock waves induce anabolic effects in osteoporotic rats. J Orthop Res 2013;31:768–775.

83 Koolen MKE, Kruyt MC, Zadpoor AA, Oner FC, Weinans H, van der Jagt OP: Optimization of screw fixation in rat bone with extracorporeal shock waves. J Orthop Res 2017, Epub ahead of print.

84 Notarnicola A, Tamma R, Moretti L, Fiore A, Vicenti G, Zallone A, Moretti B: Effects of radial shock waves therapy on osteoblasts activities. Musculoskelet Surg 2012;96:183–189.

85 Gollwitzer H, Gloeck T, Roessner M, Langer R, Horn C, Gerdesmeyer L, Diehl P: Radial extracorporeal shock wave therapy (RESWT) induces new bone formation in vivo: Results of an animal study in rabbits. Ultrasound Med Biol 2013;39:126–133.

86 Devas MB: Stress fractures of the tibia in athletes or shin soreness. J Bone Joint Surg Br 1958;40:227–239.

87 Saunier J, Chapurlat R: Stress fracture in athletes. Joint Bone Spine 2017;pii:S1297-319X(17)30097-0.

88 Cook JL, Purdam CR: Is tendon pathology a continuum? A pathology model to explain the clinical presentation of load-induced tendinopathy. Br J Sports Med 2009;43:409.

89 Hotzinger A, Radelb L, Lauber US, Lauber H, Platzekc P, Ludwig J: MRI-guided SWT of multiple stress fractures of the tibia. Transactions of the ISMST 2nd International ISMST Congress, London, 1999.

90 Leal C, Herrera JM, Murillo M, Duran R, Reyes OE, Lopez JC: ESWT in high performance athletes with tibial stress fractures. Transactions of the ISMST 5th International ISMST Congress, Winterthur, 2002.

91 Audain R, Alvarez Y, Audain R, Perez N, Barrios G: Focused shockwaves in the treatment and prevention of tibial stress fractures in athletes. in: Transactions of the ISMST 15th International ISMST Congress, Cartagena, 2012.

92 Gordon R, Lynagh L: ESWT treatment of stress fractures. Transactions of the ISMST 5th International ISMST Congress, Winterthur, 2002.

93 Taki M, Iwata O, Shiono M, Kimura M, Takagishi K: Extracorporeal shock wave therapy for resistant stress fracture in athletes: a report of 5 cases. Am J Sports Med 2007;35:1188–1192.

94 Moretti B, Notarnicola A, Garofalo R, Moretti L, Patella S, Marlinghaus E, Patella V: Shock waves in the treatment of stress fractures. Ultrasound Med Biol 2009;35:1042–1049.

Priv.-Doz. Dr. Rainer Mittermayr
AUVA Trauma Center Meidling
Kundratstrasse 37
AT–1120 Vienna (Austria)
E-Mail rainer.mittermayr@auva.at

Wang C-J, Schaden W, Ko J-Y (eds): Shockwave Medicine.
Transl Res Biomed. Basel, Karger, 2018, vol 6, pp 64–69 (DOI: 10.1159/000485062)

Local and Systemic Effects of Extracorporeal Shockwave Therapy on Bone

Sergio Russo · Valeria Servodidio · Giuseppe Mosillo · Francesco Sadile

Department of Public Health, School of Medicine and Surgery, University of Naples Federico II, Naples, Italy

Abstract

The authors examine the mechanisms of action of shockwave (SW) in bone regeneration and report the latest studies published in recent years. They all show a local or even synergistic effect of SW on bone. SW therapy is actually used for the treatment of orthopedic diseases – to induce a callus formation in non-union of long bone fractures. Currently, the method is valid and also used in cardiology and neurology as well as in the aesthetic department. This wide scope finds its rationale in the knowledge of the mechanisms of action of SW. These are no longer considered a means to destroy a tissue but as a form of energy with which to stimulate the metabolic response of the target cells by inducing in this way a metabolic reparative response. Our study shows a systemic effect of SW on bone. How could the SW therapy, applied in a single area, induce a systemic effect? This is still the subject of many studies. Our hypotheses are linked to the possibility by SW to induce the activation of a signal transmitter is able to reach other body site far from the area of application inducing a systemic effect that could also be centrally regulated. © 2018 S. Karger AG, Basel

Introduction

From the 1980s, the therapeutic use of shockwaves (SW) has been gradually extending from urology to the orthopedic and rehabilitation field [1]. Afterwards, SW therapy has been used for the treatment of other orthopedic diseases, accelerating bone healing, callus formation, and delayed or non-union of long bone fractures [2–4]. In addition, SW has been shown to promote the regeneration of alveolar bone in a rodent model of periodontitis [5]. This therapy is considered a safe and highly versatile tool to enhance the time of tissue regeneration, in particular on tendon and muscle tissues, also showing immediate antalgic and anti-inflammatory effects [6].

Currently the method is valid also used in plastic surgery, cardiology, neurology, and dermatology as well as in aesthetic medicine. This wide scope finds its rationale in the knowledge of the mechanisms of action of SW. These are no longer considered a means to destroy a tissue (kidney stones) but as a form of energy with which to stimulate the metabolic response of the target cells by inducing in this way a metabolic reparative response.

Our study shows a new therapeutic application in the treatment of osteoporosis. The drug therapies actually used are not yet able to eradicate the problem, but they are able to only mitigate it and forcing patients to take antiosteoporotic and antiinflammatory drugs for life. How could the SW therapy, applied in a single area, induce a systemic effect? This is still the subject of studies. Our hypothesis, on which we are now working, are linked to the possibility by SW to induce the activation of a signal transmitter (gasous or hormonal), which is able to reach other body sites far from the area of application inducing a systemic effect that could also be centrally regulated.

Clinical Studies on Bone Effects of SW

In 1991, Valchanou and Michailov published the first article on the clinical effect of SW on bone tissues [7]. They demonstrated for the first time that the high-energy SWs were able to consolidate the torpid fractures of the long bones and this effect was related to cortico medullary lesions produced by the same SW into the fracture areas. This effect is an expression of a local mechanism of action and mainly due to the mechanical effects of SWs. These data were later also confirmed by other experimental work conducted by other authors both in vitro and in vivo [8, 9]. The use of SW in the pseudarthrosis therapy was started from these early clinical and experimental studies. In the following years, several authors have also observed and described local biochemical effects. They are due to the increase or activation of particular enzymatic chains such as those of the nitric oxide synthase [10]. These observations have led to the use of SW in bone edema therapy, bone necrosis, and other ischemic diseases involving different tissues from bone [11, 12].

Preclinical Studies on Bone Effects of SWs

The biological effects of the SW therapy in bone have been recently examined [13]. Indeed, Van der Jagt et al. [14] have demonstrated that a single application of SW has a light beneficial effect in a rat model of ovariectomy-induced osteoporosis, increasing trabecular bone volume and reducing bone loss. Additionally, this research group showed that a single application on tibia induces anabolic effects in cortical bone in normal and osteoporotic rats, especially when SW treatment was combined with antiresorptive alendronate therapy [15, 16]. However, the bone biochemical mechanisms underlying the anti-osteoporotic effects of SWs are still overlooked.

Osteoporosis is a metabolic skeletal disease characterized by an imbalance between osteoclast-mediated bone resorption and osteoblast-mediated bone formation. Postmenopausal hypoestrogenism generates the events inducing osteoporosis: massive bone resorption leading to loss of bone mass and architecture and failure to replace lost bone due to the reduced bone formation [17]. The ovariectomized (OVX) rat is a standard preclinical model in the development of anti-osteoporosis therapies, reproducing molecular and biochemical alterations of this disease [18, 19].

In our recent study, we demonstrated clinical and theoretical implications about the field of osteoporosis therapy and the systemic effects of SW therapy [20]. The study was conducted in ovarectomized osteoporotic rats. In these animals, in which the femoral right was treated with SW, we obtained the recovery of the lost bone mass due to osteoporosis. Both bone parameters, blood and histological, show a quantitative and qualitative functional recovery in a short time (5 weeks). But even more interesting is the evidence that the effect is not local but systemic. In fact, we observed a recovery of the lost bone mass in the contralateral limb and vertebra. In this study, we valued the effect of SWs alone or the effect of SWs in combination with raloxifene (RAL) on bone loss in this animal model of osteoporosis. Sixteen weeks after surgery, osteoporotic rats were treated for 5 times (one session weekly) with SW at the antero-lateral side of the right hind leg, at 3 Hz (Duolith®, EFD of 0.33 mJ/mm² or with RAL (5 mg/kg/die, per os). SW, alone or combined with RAL, prevented femur weight reduction and resulted in the deterioration of trabecular microarchitecture both in SW-treated and -untreated femur and vertebrae, improving bone mineral density, altered by OVX. SW remarkably ameliorated the thickness and organization of metaphyseal trabeculae, and notably their combination with RAL restored trabecular structure as that of control rats without difference between SW-treated and contralateral limb, supporting the hypothesis of a systemic effect of SW therapy.

Several key factors regulate the balance between bone resorption and formation. Osteoclastogenesis is stimulated by the binding between RANK and RANKL, expressed in osteoclast progenitor cells and osteoblasts respectively. RANKL-RANK interaction is prevented by OPG, the natural RANKL inhibitor. Our data show that SW increased serum and bone OPG/RANKL ratio, thereby showing a reduction of the osteoclastogenic process. In tibiae of osteoporotic rats, SW therapy alone or associated with RAL significantly reduced TNF-α levels and cathepsin k, indicating also the inhibition of osteoclast activity, while significantly increasing the runt-related transcription factor 2 and bone morphogenetic-2 expression, suggesting an increase in osteoblastogenic activity. Among all bone morphogenetic protein members of transforming growth factor superfamily, Bmp2 has been recognized as a pivotal signal in regulating osteoblastogenesis [21]. Using transgenic mice it has shown that Bmp2 plays a pivotal role in preserving the regenerative capacity of bone, and proteosome inhibitors enhancing Bmp2 expression may have an effect on bone tissue for their potential anabolic effects [22, 23]. The involvement of the NO pathway in SW osteogenic effect was previously reported in vitro, indicating an increase in Bmp2 and RUNX2 transcription in marrow stromal cells of hips with osteonecrosis [24].

It is well known that tumor necrosis factor α (TNF-α) increases osteoclast lifespan inducing anti-apoptotic activity on these cells, and synergistically with RANKL, stimulates osteoclast formation and activity, inducing osteoclast and bone resorption both directly and increasing the sensitivity of maturing osteoclasts to RANKL [25, 26]. Cathepsin k is a prominent lysosomal cysteine protease that is involved in the bone degradation of extracellular matrix proteins, such as elastin and collagen [27]. Recently, it has been demonstrated that two cathepsin K inhibitors, odanacatib and ONO-5334, prevented bone loss in osteoporotic rabbits, and to date, they are currently used in clinical development, as new pharmacological therapy for osteoporosis [28].

In our experimental conditions, bone marrow obtained from tibiae of osteoporotic animals, SW reduced peroxisome proliferators actived receptors (PPAR)-α and adiponectin transcription, indicating a shift of mesenchymal cells toward osteoblastogenesis. Interestingly, we observed that SW not only markedly increased the transcription of RUNX2, a principal transcription factor for bone osteogenesis but also reduced PPARγ in bone marrow, indicating the shift of mesenchymal cells from adipogenesis to osteogenesis [29, 30].

All these data indicated SW therapy as an innovative strategy to limit the hypoestrogenic bone loss, pointing out to the fact that the mechanisms of SW effects can be a result of the increase of bone formation and the reduction of bone resorption.

Conclusions

SW therapy can be considered an effective, convenient, and safe noninvasive therapeutic modality, also having the potential to replace surgery with no surgical risks in many orthopedic disorders. It may also represent an innovative strategy to limit the progression of osteoporosis [7, 20, 31]. To date, no one had ever hypothesized the possibility of systemic metabolic effects like it is demonstrated by our study on animal model of osteoporosis.

The possibility of developing metabolic effects in the same tissue type but in nontreated areas, far from the application point, represents a total upheaval of those effects that are related to the knowledge of the mechanisms of action of the SWs; it necessarily leads us to reconsider those that constitute the current knowledge of the physiology of the bone tissue.

The results of the study described lead us to very suggestive interpretations. The stimulation of specific peripheral tissue through the SWs would lead to a modulation of the activity of metabolic and hormonal mediators, which may induce systemic effects and could not be ruled out as a central control mediated by these "SW-induced messengers." This mechanism hypothesized for bone tissue may also be present for other organs and tissues. It follows that could be susceptible to SW therapy also other systemic diseases that affect different districts other than the skelet.

Acknowledgment

A special thanks to Professor Rosaria Meli of the Pharmacy Department of the University Federico II of Naples for her valuable contribution to the drafting of this article.

References

1 Chaussy C, Schmiedt E, Jocham D, Brendel W, Forssmann B, Walther V: First clinical experience with extracorporeally induced destruction of kidney stones by shock waves. J Urol 1982;127:417–420.

2 Wang CJ, Yang KD, Ko JY, Huang CC, Huang HY, Wang FS: The effects of shockwave on bone healing and systemic concentrations of nitric oxide (NO), TGF-beta1, VGF and BMP-2 in long bone nonunions. Nitric Oxide 2009;20:298–303.

3 Narasaki K, Shimizu H, Beppu M, Aoki H, Takagi M, Takashi M: Effect of extracorporeal shock waves on callus formation during bone lengthening. J Orthop Sci 2003;8:474–481.

4 Wang CJ, Chen HS, Chen CE, Yang KD: Treatment of nonunions of long bone fractures with shock waves. Clin Orthop Relat Res 2001:95–101.

5 Sathishkumar S, Meka A, Dawson D, House N, Schaden W, Novak MJ, Ebersole JL, Kesavalu L: Extracorporeal shock wave therapy induces alveolar bone regeneration. J Dent Res 2008;87:687–691.

6 Mariotto S, de Prati AC, Cavalieri E, Amelio E, Marlinghaus E, Suzuki H: Extracorporeal shock wave therapy in inflammatory diseases: molecular mechanism that triggers anti-inflammatory action. Curr Med Chem 2009;16:2366–2372.

7 Valchanou VD, Michailov P: High energy shock waves in the treatment of delayed and nonunion of fractures. Int Orthop 1991;15:181–184.

8 Delius M, Draenert K, Al Diek Y, Draenert Y: Biological effects of shock waves: in vivo effect of high energy pulses on rabbit bone. Ultrasound Med Biol 1995;21:1219–1225.

9 Kaulesar Sukul DM, Johannes EJ, Pierik EG, van Eijck GJ, Kristelijn MJ: The effect of high energy shock waves focused on cortical bone: an in vitro study. J Surg Res 1993;54:46–51.

10 Mariotto S, Cavalieri E, Amelio E, Ciampa AR, de Prati AC, Marlinghaus E, Russo S, Suzuki H: Extracorporeal shock waves: from lithotripsy to anti-inflammatory action by no production. Nitric Oxide 2005;12:89–96.

11 Wang W, Liu H, Song M, Fang W, Yuan F: Clinical effect of cardiac shock wave therapy on myocardial ischemia in patients with ischemic heart failure. J Cardiovasc Pharmacol Ther 2016;21:381–387.

12 Di Meglio F, Nurzynska D, Castaldo C, Miraglia R, Romano V, De Angelis A, Piegari E, Russo S, Montagnani S: Cardiac shock wave therapy: assessment of safety and new insights into mechanisms of tissue regeneration. J Cell Mol Med 2012;16:936–942.

13 Cheng JH, Wang CJ: Biological mechanism of shockwave in bone. Int J Surg 2015;24:143–146.

14 van der Jagt OP, van der Linden JC, Schaden W, van Schie HT, Piscaer TM, Verhaar JA, Weinans H, Waarsing JH: Unfocused extracorporeal shock wave therapy as potential treatment for osteoporosis. J Orthop Res 2009;27:1528–1533.

15 van der Jagt OP, Piscaer TM, Schaden W, Li J, Kops N, Jahr H, van der Linden JC, Waarsing JH, Verhaar JA, de Jong M, Weinans H: Unfocused extracorporeal shock waves induce anabolic effects in rat bone. J Bone Joint Surg Am 2011;93:38–48.

16 van der Jagt OP, Waarsing JH, Kops N, Schaden W, Jahr H, Verhaar JA, Weinans H: Unfocused extracorporeal shock waves induce anabolic effects in osteoporotic rats. J Orthop Res 2013;31:768–775.

17 Raisz LG: Pathogenesis of osteoporosis: concepts, conflicts, and prospects. J Clin Invest 2005;115:3318–3325.

18 Lelovas PP, Xanthos TT, Thoma SE, Lyritis GP, Dontas IA: The laboratory rat as an animal model for osteoporosis research. Comp Med 2008;58:424–430.

19 Meli R, Pacilio M, Raso GM, Esposito E, Coppola A, Nasti A, Di Carlo C, Nappi C, Di Carlo R: Estrogen and raloxifene modulate leptin and its receptor in hypothalamus and adipose tissue from ovariectomized rats. Endocrinology 2004;145:3115–3121.

20 Lama A, Santoro A, Corrado B, Pirozzi C, Paciello O, Pagano TB, Russo S, Calignano A, Mattace Raso G, Meli R: Extracorporeal shock waves alone or combined with raloxifene promote bone formation and suppress resorption in ovariectomized rats. PLoS One 2017;12:e0171276.

21 Wagner N, Weyhersmuller A, Blauth A, Schuhmann T, Heckmann M, Krohne G, Samakovlis C: The drosophila lem-domain protein man1 antagonizes bmp signaling at the neuromuscular junction and the wing crossveins. Dev Biol 2010;339:1–13.

22 Rosen V: BMP2 signaling in bone development and repair. Cytokine Growth Factor Rev 2009;20:475–480.

23 Appelman-Dijkstra NM, Papapoulos SE: Novel approaches to the treatment of osteoporosis. Best Pract Res Clin Endocrinol Metab 2014;28:843–857.

24 Yin TC, Wang CJ, Yang KD, Wang FS, Sun YC: Shockwaves enhance the osteogenetic gene expression in marrow stromal cells from hips with osteonecrosis. Chang Gung Med J 2011;34:367–374.

25 Teitelbaum SL: Osteoclasts: What do they do and how do they do it? Am J Pathol 2007;170:427–435.

26 Weitzmann MN, Pacifici R: The role of t lymphocytes in bone metabolism. Immunol Rev 2005;208:154–168.

27 Troen BR: The role of cathepsin K in normal bone resorption. Drug News Perspect 2004;17:19–28.

28 Pietschmann P: Cathepsin K inhibitors: emerging treatment options for osteoporosis. Wien Med Wochenschr 2015;165:47.

29 Vimalraj S, Arumugam B, Miranda PJ, Selvamurugan N: Runx2: Structure, function, and phosphorylation in osteoblast differentiation. Int J Biol Macromol 2015;78:202–208.

30 Sadie-Van Gijsen H, Crowther NJ, Hough FS, Ferris WF: The interrelationship between bone and fat: from cellular see-saw to endocrine reciprocity. Cell Mol Life Sci 2013;70:2331–2349.

31 MC DA, Frairia R, Romeo P, Amelio E, Berta L, Bosco V, Gigliotti S, Guerra C, Messina S, Messuri L, Moretti B, Notarnicola A, Maccagnano G, Russo S, Saggini R, Vulpiani MC, Buselli P: Extracorporeal shockwaves as regenerative therapy in orthopedic traumatology: a narrative review from basic research to clinical practice. J Biol Regul Homeost Agents 2016;30:323–332.

Dr. Prof. Sergio Russo
Department of Orthopedic Surgery, University Federico II of Naples
Via S. Pansini 5
IT–80131 Naples (Italy)
E-Mail serghiey@hotmail.com

Wang C-J, Schaden W, Ko J-Y (eds): Shockwave Medicine.
Transl Res Biomed. Basel, Karger, 2018, vol 6, pp 70–86 (DOI: 10.1159/000485063)

Extracorporeal Shockwave Therapy and Sports-Related Injuries

Carlos Leal · Edmundo Berumen · Arnold Fernandez ·
Susana Bucci · Andres Castillo

Fenway Medical Shockwave Medicine Unit, Postgrado de Ortopedia y Traumatologia, Facultad de Medicina, Universidad El Bosque, Bogota, DC, Colombia

Abstract

Sports and physical activities have become a key in the daily living of the present century. Well-being and fitness are common words in modern language. High performance comes with a quote of physiological demands that have to go beyond normal to be competitive. These factors have increased the number of injuries both from acute trauma and from biomechanical overuse. Treatments are often related to pain and load control, medication and physical therapy. Results are usually positive, but in certain number of patients, the sports-related condition becomes chronic. At this stage, the biological physiopathology is more of a disease than a mechanical injury. It comes with changes in blood supply and vascularity, and a lack of healing capacity of the tissues. The poor migration and differentiation of cells result in lower metabolic activity that usually develops a biomechanical derangement of function. Treatment changes target: revascularization and cell repopulation. This is the case of chronic tendinopathies and stress fractures. Conventional treatments may not be enough in certain cases, and surgical procedures are commonly recommended. However, they are associated with complications, long periods of recovery, pain, and high costs. Extracorporeal shockwave treatments (ESWT) have provided a noninvasive therapeutic tool that enhances healing by means of revascularization and tissue stimulation through mechanotransduction. Sports injuries are the perfect scenario for ESWT. Chronic conditions such as tendinopathies or stress fractures are common in sports overuse syndromes, and failure to succeed with conventional medical treatments usually ends with the indication for surgical procedures. ESWT has replaced the need for surgery, and become a common treatment with no complications and great pain control and function recovery. Standardization of techniques and protocols, larger randomized control trials, and new research lines for basic sciences and understanding biological pathways of mechanotransduction are the main goals of the International Society for Medical Shockwave Treatments. These efforts will strengthen the power of ESWT for sports medicine and other medical specialties.

Extracorporeal Shockwave Therapy (ESWT) has been used for different medical conditions in the past 3 decades [1–5]. The development of the technology for treating kidney stones in the 1980s by high-energy-focused ultrasound marked a new medical treatment concept that changed long established paradigms in medicine and urology. Seeking for similar results in calcified tendinopathies led to basic and clinical research efforts to find a possible application in musculoskeletal diseases. The results in terms of removal of calcified tissues were not as positive as in urology, but the key finding was the improvement of pain control and function that remained in time. The next step was the application of ESWT in chronic tendinopathies, a common clinical problem in sports medicine. The preliminary results were encouraging, and the development of protocols for specific tendinopathies led to solid reports and publications, a growth in technology, and a notable interest in understanding the molecular basic sciences of mechanotransduction [6].

Bone applications of shockwave medicine were the first objective of many research groups around the world. The fascinating ESWT stimulation of bone turnover with osteo-inductive responses and neo-vasculogenesis became a solid therapeutic tool for delayed unions and pseudoarthrosis. This effect was also used in the treatment of early stages of stress fractures, avascular necrosis femoral head, and subchondral bone stimulation in osteoarthritis and Osteochondritis Dissecans [4, 7, 8].

Many basic science research studies were performed in skin models, as fibroblasts react very positively to mechanotransductional stimulation. The number of capillaries and pericites in the dermis is also a good target for high-energy ultrasound. In the early days, infectious diseases and ulcers were a contraindication to the use of ESWT in fractures. However, some patients treated for delayed unions that had open wounds showed an unexpected skin healing positive response. The basic studies showed a significant increase in cell migration and differentiation, as well as an improvement in vascularity, that led to the first reports of the use of ESWT in diabetic foot ulcers, delayed wound closure, and complex combat wounds. The positive metabolic effect of ESTW has also been used in esthetic medicine to stimulate collagen production in striae, cellulitis, keloids, and hypertrophic scars [9].

The evolution of ESWT for tendon, bone, and skin healing disorders have opened a wide research window that took the scientific societies and investigation groups to deeply explore the biological principles of mechanotransduction [10]. Even though there are many processes yet to be discovered, we know that ESWT stimulate the natural communication pathways between cells by increasing the production of intercellular messengers such as growth factors, specific proteins, and enzymes [9]. Mechanotransduction stimulates and enhances the gene expression systems of cells, increasing their regular functions and invigorating their capacity to call for undifferentiated cell migration and differentiation. Many cell metabolism biomarkers such as Wingless Proteins (WNT), Von Willebrandt factor, TGFs and NOs have been shown to increase in vitro as a response to ESWT [11]. Clinical applications have a better ground day by day, and ESWT technology has become a therapeutic tool that gains power for many different applications in medicine.

The initial knowledge of ESWT based its tissue responses on mechanical and not biological effects. The biomechanical principle of physically impacting tissues with different acoustic impedances was the basic assumption for kidney stone destruction and bone healing. This was the reason why the first tendinopathies treated with ESWT were associated with calcifications, such as shoulder rotator cuff tendinopathies, heel spurs, or tennis elbows. High-focused energies were used, and results showed improvement of pain and function, but irrelevant bone spur dissolution.

The use of lower energies and the introduction of radial pressure waves devices in 2000, as well as the advances in the investigation of biological principles of mechanotransduction, led to understanding both the physiopathology of chronic tendinopathies and the role of ESWT in their treatment [12]. Tendon and ligament healing disturbances with low vascularity, poor migration, and differentiation of tenocytes, decreased protein or collagen production and low cellularity are the reasons for chronicity in tendinopathies. ESWT has proved a positive stimulation effect on all of these fields.

Sports injuries are common, not only in high-performance athletes but also in the regular population that workout and exercise. The most common lesions are muscle pulls, followed by ligament and tendinous sprains. The majority of lesions are mild, and the large articular injuries are relatively easy to diagnose. An anterior cruciate ligament rupture or a torn meniscus is usually diagnosed within the first hours or days, and the treatment planned accordingly. Acute lesions are common, and are usually treated by well-known protocols that have high rate of good results [13]. However, the incapacity to heal in a regular manner creates chronicity of lesions, especially in ligaments and tendons.

Pain control using a noninvasive system that does not involve radiation, anesthesia, or medications has always been an attractive feature of ESWT in sports medicine. The first time ESWT was used in an Olympic team was by Dr. Lohrer et al. [1] in Atlanta 1996 Olympic games. He moved their heavy focused device from Germany to the United States to manage pain control in the German Olympic Team. They reported great results but emphasized the need for lighter and more affordable devices. By the end of the century, another urological development was used for ESWT technology. The use of endoscopic ballistic pressure lithotripters allowed kidney stone destruction similar to the high-energy extracorporeal devices. ESWT companies started to explore the use of ballistic pressure waves for musculoskeletal pain, and changed the whole picture of the technology and its applications.

The most common sports-related lesions that have an ESWT indication are chronic tendinopathies and stress fractures. Both Shin splint syndrome and stress fractures are common sports-related pathologies and are well treated with ESWT.

It is safe to state that most of all sports-related tendinopathies share a similar physiopathology and treatment, and account for more than 82% of the applications of

ESWT in the world. The most common are shoulder chronic tendinopathies with or without calcifications, lateral epicondylopathy, and patellar tendinopathy. Insertional tendinitis evolves into a chronic stage with lack of differentiated tissue count for approximately 25% of all acute tendon-related pain conditions.

Most of the anatomical sites are subcutaneous, painful, and palpable. This means that shockwave focusing is not an issue, and the energy absorbed by intermediate elements is scarce. The development of ballistic pressure wave devices that deliver low energies with more affordable devices generated a worldwide boost of applications of what was called Radial Shockwave Therapies (RSWT). Many professional sports teams, national Olympic sports groups, and rehabilitation centers opened shockwave units. Non-physicians run many of these centers, a fact that also created difficulties in adequate indications for treatment. ESWT is a medical-based procedure that requires a diagnostic protocol that must be performed by a medical doctor.

The International Society for Medical Shockwave Treatments (ISMST) includes both tendon and bone sports-related conditions within their approved standard indications.

Tendon and Ligament
ISMST Approved indications
– Lateral epicondylopathy of the elbow
– Patellar tendinopathy
– Achilles tendinopathy
– Plantar fasciitis, with or without heel spur
Common empirically tested clinical uses
– Rotator cuff tendinopathy without calcification
– Medial epicondylopathy of the elbow
– Adductor tendinopathy syndrome
– Pes-Anserinus tendinopathy syndrome
– Peroneal tendinopathy
– Foot and ankle tendinopathies
Bone
ISMST-approved indications
– Stress fractures
– Avascular bone necrosis without articular derangement
– Osteochondritis Dissecans without articular derangement
Common empirically tested clinical uses
– Bone marrow edema
– Osgood Schlatter Disease: apophysitis of the anterior tibial tubercle
– Tibial stress syndrome (shin splint)
Bursae
ISMST-approved indications
– Greater trochanteric pain syndrome

Muscle

Common empirically tested clinical uses

– Muscle sprain without discontinuity

ESWT for Sports-Related Tendinopathies

Most patients and athletes that are seen in the orthopedic or sports medicine units because of tendon-related pain already have a chronic degenerative tendon disease. This noninflammatory disorder is often mislabeled and mistreated as tendinitis, an acute condition that usually responds properly to rest, medication, and physical therapy [14]. Tendon lesions can be secondary to acute trauma such as a rupture or a laceration, or by repetitive loading or overuse injuries. According to Dr. Woodwell and Dr. Cherry [15] and the US Department of Labor, these chronic tendinopathies represent a serious health problem, being as high as 7% of all physician visits in the United States and affect 100 of 1,000 of workers causing a significant loss of work time [15, 16].

Physicians must recognize the physiopathology of every specific injury and be aware of the anatomic, biomechanic, and biologic background of every case. The natural history is a gradually localized increasing pain related to exercise and sports. Most patients report engagement in a new physical activity, sport, or training routine with different mechanical demands. Patients describe the pain as a sharp localized stabbing sensation that evolves into an uncomfortable dull ache after activities and rest.

Examination should include motion analysis, a thorough inspection assessing points of tenderness, swelling, erythema, and symmetry with the contralateral limb. Every tendon has well-described examination maneuvers and validated tests that are positive as pain appears under loading.

Diagnostic images include a plain X-ray test, a simple diagnostic tool that may overrule deeper problems such as arthritis, stress fractures, bone cysts, or tumors. It is a nonexpensive, noninvasive image, and no patient should be treated for any tendinopathy without at least a plain X-ray test. Ultrasound is very useful to determine the active motion of tendons and joints, and to document the echographic patterns of degenerative areas. MRI is extremely useful to diagnose and measure degenerative areas and to precisely define any differential diagnosis. It is especially useful in shoulder, patellar, and Achilles tendinopathies [17–19].

Treatment for acute tendinitis is based on pain control and physical therapy [20]. Anti-inflammatory medication, rest, and therapy usually work well. Injected corticosteroids (CSs) can also relieve pain but must be used with caution. In chronic conditions, where the inflammatory process is lower and the hipovascular and degenerative areas are larger, treatment must be based upon revascularization and recovery of the biological and biomechanical homeostasis of the tendon. Some strategies include orthobiologic injections, tendon punctures, laser, ultrasound, radiofrequency, or electric stimulation. Surgery provides the largest possible vascular stimulation and removal of degenerative tissue due to direct trauma to the tendon. However, it has a larger number of possible complications that include infection, a long recovery, high cost, and the need for anesthesia. ESWT and mechanotransductional stimulation of the degenerative areas of tendon has proven similar results to surgery, without complications [21, 22].

Major research of the effects of ESWT on biologic tissues has been done on tendinopathies [23, 24]. The 2 major effects we seek for in these overused conditions are pain control and tissue regeneration. Pain relief can be related to a hyper-stimulation analgesic effect, as well as a block of ascendant neurological signals to the central nervous system. Some animal studies have shown an influence of ESWT on Substance P, as well as the dorsal root expression of calcitonin gene-related peptide CGRP, 2 documented markers of musculoskeletal pain [25, 26]. There are controversial points of view on the pain relief pathways of ESWT that are under research. What is clear is that there is a clinical effect of pain relief in every report of ESWT and tendinopathies in the literature, and patients reveal an analgesic effect with the use of shockwaves.

But tissue regeneration is probably the major goal in chronic tendinopathies treated with ESWT. There are many ways of controlling pain, but targeting the primary cause of the disease is the keystone for tendinopathy treatments. ESWT mechanically stimulates the cytoskeleton creating cell responses such as increased protein synthesis, and intercellular messenger releases [5, 27]. We know this bimolecular process as mechanotransduction.

Dr. Vetrano et al. [28] showed that ESWT can promote proliferation and collagen production in human tenocytes and Dr. Han et al. [29] proved that ESWT in vitro decreased the expression of matrix metalloproteinases MMPs and Interleukins associated with chronic tendinopathies. Studies from Dr. Wang et al. [27], Dr. Bosch et al. [30] and Dr. Hsu et al. [31] have proven an increase in matrix turnover and vascularization of the tendon-bone junction in animal models. Dr. Brañes et al. [32] showed a clear pattern of adult type of revascularization in rotator cuff samples of human shoulders previously treated with ESWT, different from inflammatory neo-angiogenesis, proving a clear effect of *capillaryzation* from vascular pericites.

However, the clinical validation of the effects of ESWT in vivo is by far more difficult. The major outcome is pain relief and functional scales that are extremely variable. Most clinical studies provide low power evidence, and the protocols, devices, energies, and evaluations are extremely variable. The validation of safety is always positive, showing no serious complications or adverse effects in the use of ESWT in tendinopathies. The efficacy, as compared to other methods is at least similar or better. The general parameters and guidelines for application are provided by the scientific societies and the industry, but operators are free to use the technology in the way they desire, as it is a noninvasive and safe procedure. This has led to a number of scattered data that is difficult to compare. The common finding in the literature is that ESWT is a safe and effective therapeutic method for sports-related tendinopathies,

Table 1. There is a high variability of ESWT protocols for chronic tendinopathies

ESWT	Energy	Number of shockwaves per session	Number of sessions	Application interval, days
Focused electrohydraulic, mJ/mm^2	0.01–0.15	2,000	2–3	7–21
Focused electromagnetic or Piezoelectric, mJ/mm^2	0.25–0.40	2,000	2–3	7–21
Radial, Bar	2–4	4,000	3–4	7–21

The summarized data from the literature and the recommendations from recognized experts and approved independent shockwave centers from all over the world, provide the common guidelines of use from the ISMST. These protocols and guidelines are common for all tendinopathies, with local anatomical variability.

and that the increased number of well-designed studies and basic science research efforts will lead to a better design of treatment protocols.

The treatment protocols have been extremely variable due to the different guidelines provided from the industry and the literature. The energy level, the number of sessions, the frequency of shockwaves, and the type of applicator are reported in many different ways. This creates confusion for users, patients, and insurers. The bright side is that despite the differences, outcomes are always positive, or at least never negative or with adverse effects or complications. The dark side is that a safe and effective technology to treat tendinopathies such as ESWT does not have a consented dosage or protocol to be recommended with every device available in the market (Fig. 1).

The ISMST has recommended the protocols with the highest levels of evidence in the statements of approved applications. The protocols are similar for all tendinopathies, and the variations reside on the site and direction of the shockwaves. The ISMST recommended protocols for focused and radial ESWT in chronic tendinopathies is presented in Table 1.

Lesions of the rotator cuff and the biceps tendon are the most common sports-related shoulder tendinopathies. Calcified tendinopathies of the rotator cuff are related to chronicity, and are often not associated with exercise or sports. Most of the chronic shoulder tendinopathies have a multifactorial genesis that cannot be just defined as an insertional tendinopathy. There is a clear mechanical influence of the passage of the rotator cuff though a bony narrow tunnel that closes dynamically as the joint flexes or elevates. There is a demonstrated lack of vascularity associated with calcified tendinopathies.

A common finding on the MRI tests is a partial tear of the rotator cuff, associated with subacromial impingement or subacromial bursitis. All of these conditions are treated nonsurgically, and have a potential indication for ESWT. The literature re-

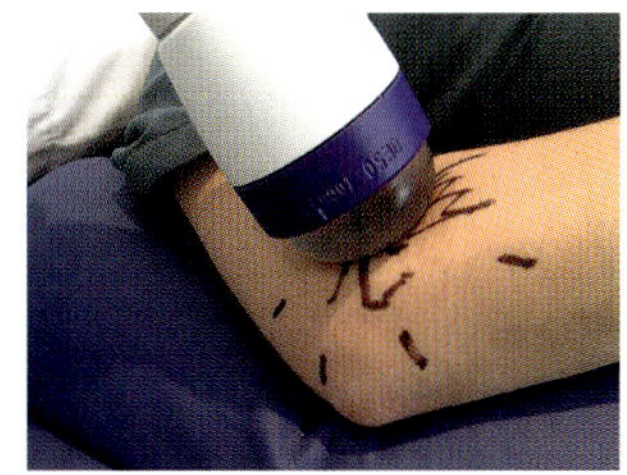

Fig. 1. The use of extracorporeal shockwaves requires a proper placement of the applicator, in areas where mechanotransduction will create a biological positive effect. In most tendinopathies, it must be focused on the bone-tendon junction, not necessarily in the most painful area. It always requires marking the treatment area, and the use of an echo transmitting gel interphase. Number of shockwaves, levels of energy, and number of sessions are similar to all insertional tendinopathies.

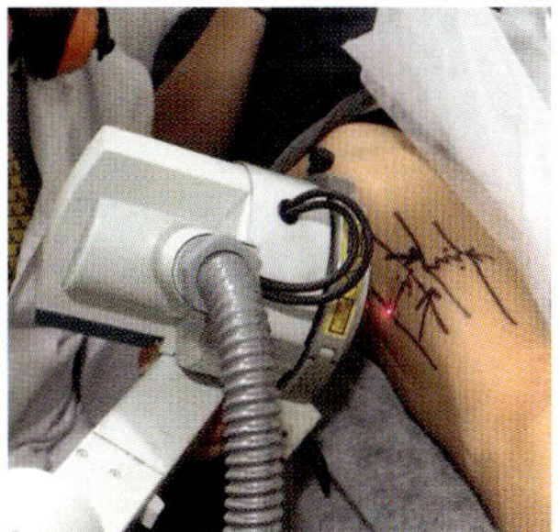

Fig. 2. Even though there are reports of good results with lower energies, the best outcomes for the extracorporeal treatment of stress fractures are obtained with high-energy-focused devices. One session under sedation, under fluoroscopic X-ray location, followed by a load control and specific rehabilitation program, provides the best option for this medical condition.

ports show a good control of pain in partial tears, as well as in early subacromial impingement syndromes [33]. Dr. Brañes et al. [32] reported good in vivo results in partial tears, with neovasculogenesis as a major histologic finding in biopsies. Dr. Galasso et al. [34] reported significantly better results with ESWT in noncalcified tendinopathies. He found that the long-term results are similar to those obtained with physical therapy and muscle balance. Most of the literature relates to pain control and function, and shows good safety with similar efficacy as conventional conservative measures.

Biceps tendinopathies can also be related to sports and are usually caused by partial tears or bone spurs, or an overuse condition. ESWT has been used for these diagnoses with relatively good results.

ESWT for calcified shoulder tendinopathies has shown good evidence of positive results in time. Dr. Wang et al. [5], Dr. Gerdesmeyer et al. [35], and others have showed significantly better results than placebo in pain and function, as well as in the disappearance of the calcification. Dr. Rompe et al. [36] compared ESWT with surgery obtaining the same results, without the complications related to invasive procedures. Dr. Rebuzzi et al. [37] did the same study comparing ESWT with arthroscopy and obtained similar results.

Literature supports the use of focused high-energy devices as the treatment of choice, even though there are several publications supporting the use of radial and low-energy devices. Dr. Ioppolo et al. [38] published the largest systematic review and meta-analysis of ESWT for shoulder pathologies, and did not find a homogeneous protocol, energy level, or device recommendation. It must be noted that the most serious complications of ESWT have been reported in shoulder pathologies. There are 2 case reports of avascular necrosis of the humeral head, related to the use of high-energy ESWT. However, there are no serious reports of complications related to this application in any of the ESWT known centers in the world, that have treated 1,000 of patients for more than 2 decades. Thiele published a series of 1,800 cases with no adverse effects different that local skin bruises.

As in other tendinopathies, the use of local anesthesia has resulted in lower results and is not recommended. The diagnosis and location of the lesion is a key for the treatment of ESWT in shoulder pathologies, as the target is not palpable or visually reachable. The use of ultrasound has been well supported, as well as fluoroscopy and CT scan markers, and is a common indication for shoulder treatments.

Tennis elbow is the common name for the lateral elbow epicondylopathy, an overuse painful syndrome of the extensor muscles of the wrist and hand on the lateral humeral epicondyle. In the inflammatory tendinitis phase, treatment is based on physical therapy stretching exercises and medication. There is usually a good response with load controls and therapy, but at least 20% of patients develop a chronic painful condition that does not respond adequately to conservative measures. Surgery is indicated only in 10–12% of patients, but very few accept the procedure and prefer to give up on all pain-related physical activities.

The chronic epicondylopathy is caused in great part by the poor regenerative capacity of this particular anatomic area. The lateral epicondyle is a mass of fibrous tissue that serves as the origin of all the extensor muscles, with very few blood vessels as compared with the medial bony epicondyle. Any inflammatory condition in the lateral epicondyle will have a poor cell migration and differentiation healing response, and in about 20% of the cases, a scar tissue of angio-fibroblastic dysplasia will replace the normal tendon. This scar tissue does not have the biomechanical properties of normal tendon, and behaves as a stiff band that transfers the loads to the tendon-bone junction with no stretching, causing pain with every extensor muscle contraction.

Treatment for pain control and inflammation has a high incidence of recurrence. As in other tendinopathies, physical therapy, load control, and medication may control the chronicity of the epicondylopathy. However, rates of over 20% of patients recur and require other interventions such as ortho-biologic injections or surgery. ESWT provides a noninvasive treatment with no complications that create neovascularization and stimulation of cellularity at the tendon bone junction.

In the 1990s, the ESWT treatment of tennis elbow rose with great results and several positive publications. It even was the first indication of ESWT approved by the United States Food and Drug Administration FDA. However, several negative reports were also published and they created a cloud of doubt that moved the scientific world to understand better the role of ESWT in both tendinopathies and lateral epicondylopathy.

Dr. Vavken et al. [39] published a systematic review in 2014 that analyzed all solid evidence based medicine (EBM) studies, especially on behalf of the diverging results and outcomes as well as to point out possible conflicts according to the study-design and the use of different shockwave-devices. He found a high diversity of treatment protocols, evaluation scores, and concluding end points. He stated that the evidence in the literature supports the use of ESWT for lateral epicondylitis with symptoms beyond 3 months.

Most of the usually cited papers that do not recommend ESWT for tennis elbow have conflicting methods, statistical analysis, and conclusions. Dr. Crowther et al. [40] compared ESWT to CS injections in 93 patients in 2002, and reported 24% less success in the shockwave group. However, his main variable was short-term pain control that clearly favors injected analgesics over any treatment. Curiously, various series were published in 2002 reporting that ESWT treatment was ineffective. However, the levels of energy were lower than the recommended levels for this treatment; they used data from various centers with different devices and used local anesthesia, a clear contraindication in ESWT tendinopathy treatments. One of the most cited papers came from Dr. Speed et al. [41], with similar conclusions, but also with similar methodological problems: very low energy ESWT protocols and short-term follow-ups. Along the next 15 years, these papers have been controverted and are not a solid scientific reference to the current state of knowledge of ESWT for lateral epicondylitis.

Some authors like Dr. Buchbinder [42], Dr. Melikyan et al. [45], Dr. Chung Dr. Wiley [43], and Dr. Staples et al. [46] have concluded that there is little evidence to support the use of ESWT for the treatment of lateral epicondylitis. All these studies found that ESWT had at least some positive effects, a high safety profile with no complications, and similar or worse results when compared to conventional treatment. The common use of subtherapeutic energies is consistent in all of these publications, as well as a short-term follow up. One can conclude from all these negative results in the literature that the effects of ESWT are highly safe and moderately effective when compared in short terms with placebo, even with under-dosed energy levels. These papers did contribute to a better understanding of the effects of ESWT in insertional tendinopathies, in determining the best energy levels and protocols, as well as to read between the lines in conclusions.

Rompe performed in 2004 a placebo-controlled trial using repetitive low-energy shockwave treatment, in response to the conflicting evidence regarding ESWT for chronic tennis elbow [47]. In this well-designed trial conducted using approved protocols and energy levels, a significant benefit of ESWT when compared to sham treatment for tennis elbow, even at 3 months after intervention was reported. He had reported in 1996 similar results on an RCT [48].

Dr. Levitt presented to the American Academy of Orthopedic Surgeons his FDA approval series in 2004, showing 14% better results with ESWT than placebo for tennis elbow [49] Dr. Pettrone and Dr. McCall [50] studied 114 patients for a second FDA study, and found 61% of good results in pain control and function as compared with 29% in the placebo group. Dr. Radwan et al. [51] in 2008 compared ESWT with a percutaneous tenotomy surgery, and found nonstatistically significant better results with the surgical procedure. There was not even a single record of complications, but he stated that the results being so similar, favored a noninvasive procedure such as ESWT.

Another interesting issue is the comparison between Platelet Rich Plasma injections and ESWT. There are very few studies to determine benefits of one or the other. Dr. Ozturan et al. [52] compared in 2010 the use of autologous blood, CS injections, and ESWT for lateral epicondylitis. CS injections were better in the first weeks, blood injections in the midterm, and ESWT was the best after a 52-week follow-up. Recurrence was higher with CS injections.

Chronic patellar tendinopathy is a combination of hypo-vascularity and fibrous dysplasia that changes the biomechanical properties of the tendon and causes pain. The first reports of patellar tendinopathy treated with ESWT were done with focused devices. Dr. Vara in 2000 published a well-designed blinded RCT that showed 74% improvement in pain and function after 24 months in the ESWT-treated group [53]. Dr. Ogden et al. [54] in 2003 presented for the first time in an ISMST world meeting the results of ESWT in 11 athletes with encouraging results. Dr. Taunton in 2003 proposed ESWT as a useful complement to squat physical therapy protocols. He found an improvement of 77% in pain and function [55]. Dr. Peers et al. [56] in 2003 found 61% pain and functional improvement at 3 months in 41 athletes treated with ESWT. He also published a comparison between ESWT and surgery for patellar tendinopathy (PT), and did not find any statistically significant differences in pain or function after 24 months. He concluded that ESWT and surgery are similar in results, with a higher cost, risk, recovery time, and inconveniences of surgical procedures.

Vulpiani found improvement in pain and function of 43% after 1 month and 79% after 24 months [57]. Dr. Wang et al. [58] compared ESWT and physical therapy treatment in a controlled RCT of 50 patients. He had 90% good or excellent results in the ESWT group, as compared to 50% in the physical therapy group after 12 months. Using diagnostic ultrasound, he also reported vascular changes in the proximal insertion of the patellar tendon, and significant differences in tendon width and size in the ESWT-treated patients.

Dr. Lohrer et al. [18] in 2002 had the first report of treatment of PT with radial pressure waves. He reached 64% of good results in 50 patients, lower than those reported with focused protocols. In 2006, our group reported better results using radial pressure waves and PRP augmentation in severe PT, and recommended the use of MRI to determine the size of the degenerative defect [59]. In defects larger than 50% of the width of the proximal insertion of the tendon, the primary recommendation is surgery. The use of radial devices became popular due to good results, and for many years, evidence and reports were growing. Dr. Furia et al. [60] showed in 2013 good results in patients treated with a single session of low energy radial pressure waves, with a significant improvement in pain and function.

Dr. van der Worp [61] reported the first randomized controlled trial to compare the effectiveness of FSWT and RSWT in the treatment for patellar tendinopathy. They used a 3-session protocol with energies of 1.2 mJ/mm^2 in FSWT and 2.4 BAR in RSWT. They found a slight improvement in all patients after 14 weeks and no statistically significant difference in the effectiveness of FSWT and RSWT for treating patellar tendinopathy.

Dr. Serrano et al. [62] in 2014 reported a comparative study between radial and focused shockwaves for PT. Only radials had 67% of good and excellent results, as compared with Focused that had 88%.

The latest application reported on patellar tendinopathy was done by our group, in patients with anterior knee pain during low impact sports after a total knee arthroplasty [44]. This is a relatively common condition that causes anterior pain and discomfort in patients with total knee replacement that are physically active. Our results were good in 69% of the patients after 3 months of treatment. We found no complications and we are currently running a phase-2 case-control study.

To date there are only 2 reports in the literature showing poor results of ESWT in PT: Dr. Zwerver et al. [63] using a piezoelectric device in high-performance athletes, and Dr. Vetrano et al. [64] who compared ESWT with PRP. The first study was conducted among athletes during competition, and that may be the reason for having lower good results than others reported in the literature. In Vetrano's study, he used an unusual protocol of one session every 48–72 h that has no previous validation and is not used by other authors [64].

ESWT for sports-related tendinopathies has shown enough evidence to prove a good balance of safety and efficacy. The major recommendations are to use ESWT in tendinopathies beyond 3 months of failed treatment, avoid the use of local anesthesia, use between 2 to 3 sessions with approved protocols and devices, continue with physical therapy lesion-specific protocols, and follow up patients for more than 6 months to evaluate results. More work and research are required in order to determine precise dosage and guidelines.

ESWT for Sports-Related Bone Pathologies

Repetitive cyclic loading of bones is the most relevant etiologic factor in the genesis of sports-related bone pathologies. The fine balance between bone microdamage and remodeling marks the outcome of bone failure under repetitive loading conditions [6, 65, 66]. The most common scenario for bone failure under fatigue loading in an athlete is that of normal bone subject to abnormal or excessive loading. This condition is defined as a stress failure or stress fracture. The most common bones affected are tibia, metatarsals, fibula, navicular, pelvis, and femur [6, 67].

Stress fractures are common painful conditions in athletes, usually associated with biomechanical overloads. Low-risk stress fractures usually respond well to conservative treatments, but up to one third of the athletes may not respond, and evolve into high-risk stress fractures. Surgical stabilization may be the final treatment, but it is a highly invasive procedure with known complications. Shockwave treatments based upon the stimulation of bone turnover, osteoblast stimulation, and neovascularization by mechanotransduction have been successfully used to treat delayed unions and avascular necrosis. Since 1999, it has also been proposed in the treatment of stress fractures with excellent results and no complications.

The different studies of the biological effects of shockwaves in bone have shown a clear effect in neovascularization, periosteal stimulation, and osteo-induction [26, 68]. This final outcome is useful in the treatment of delayed unions, avascular necrosis, and stress fractures. There is a global consensus about the influence of mechanotransduction in bone turnover, with clinical use in bone-delayed unions [4, 58, 69, 70]. Using high-energy shockwaves, there is a clear stimulation of vascularity, as well as osteoblast differentiation from mesenchymal stem cells and inactive cells [71–74]. There is also an effect on periosteal cells that contribute to cell migration and the development of callus in healing impaired bone [20, 75]. The clinical use in delayed bone fracture healing and in avascular necrosis has grown both experimentally and in clinical use in the past decade [6, 7, 66, 76–79].

Until now, there are no blinded randomized control trials that can prove with level-one evidence the benefits of ESWT for bone pathologies. However, shockwave therapy is the first treatment of choice of delayed bone unions in Austria, but still its use in stress fractures requires more evidence. However, its use is biologically logical, has no complications, and the case reports have been encouraging [80].

Dr. Hotzinger et al. [81] reported the first case of stress fractures treated with ESWT at the ISMST meeting in London in 1999. He studied the role of MRI in the diagnosis of multiple stress fractures of the tibia, and treated a case with high-energy shockwaves with good results. After this first case report, we were the first group to run a clinical study on ESWT and stress fractures [82, 83]. In 2001, we performed an Experimental, Randomized, Single Blinded, Self Controlled Clinical Study with twenty six 18-year-old navy cadets with identical bilateral tibial stress fractures [84, 85]. They were all symptomatic for more than 3 months and responding poorly to conventional treatment. We evaluated a visual analogue pain scale in four conditions: rest, sports, after sports, and pressure. We followed them for a year, and kept the physical therapy protocol, and also evaluated X-rays, bone scans, and nutritional conditions.

In the most symptomatic tibia, we applied 2000 focused shockwaves in 2 sessions one week apart, at medium-high range energy of 0.1–0.27 mJ/mm^2. The treated tibias had less pain in all clinical situations, earlier than the control tibias. Both treated and control tibias were pain free after 12 months, but pain was significantly lower in shockwave-treated tibias in periods as short as 3 weeks, both during physical activity, rest, after sports, and pressure testing. We did not find major changes in bone scans, and X-rays were significantly similar. We concluded that ESWT significantly reduced pain and recovery time in high-performance athletes with tibial stress fractures [82, 83].

The ISMST has recommended the ESWT for stress fractures protocol with the highest levels of evidence, in the statements of approved applications. One session of

2,000 focused shockwaves at 0.2–0.5 mJ/mm^2 under sedation, followed by a rehabilitation program that includes load control and progressive return to sports, is the approved guideline (Fig. 2).

The use of ESWT for stress fractures is simple, noninvasive, safe, and effective treatment with better results in early stages. More controlled studies are required, but the present literature of case series and expert reports, based upon a serious solid basic science background, marks a research path with high efficiency and safety, and certainly with a clear future in the treatment of stress fractures in sports medicine.

References

1 Lohrer H, Nauck T, Korakakis V, Malliaropoulos N: Historical eswt paradigms are overcome: a narrative review. Biomed Res Int 2016;2016:1–7.

2 Notarnicola A, Moretti B: The biological effects of extracorporeal shock wave therapy (ESWT) on tendon tissue. Muscles Ligaments Tendons J 2012;2: 33–37.

3 Ogden JA, Toth-Kischkat A, Schultheiss R: Principles of shock wave therapy. Clin Orthop Relat Res 2001;387:8–17.

4 Schaden W, Fischer A, Sailler A: Extracorporeal shock wave therapy of nonunion or delayed osseous union. Clin Orthop Relat Res 2001:90–94.

5 Wang CJ: Extracorporeal shockwave therapy in musculoskeletal disorders. J Orthop Surg Res 2012;7: 11.

6 d'Agostino C, Romeo P, Lavanga V, Pisani S, Sansone V: Effectiveness of extracorporeal shock wave therapy in bone marrow edema syndrome of the hip. Rheumatol Int 2014;34:1513–1518.

7 Wang CJ, Wang FS, Huang CC, Yang KD, Weng LH, Huang HY: Treatment for osteonecrosis of the femoral head: comparison of extracorporeal shock waves with core decompression and bone-grafting. J Bone Joint Surg Am 2005;87:2380–2387.

8 Wang CJ, Wang FS, Ko JY, Huang HY, Chen CJ, Sun YC, Yang YJ: Extracorporeal shockwave therapy shows regeneration in hip necrosis. Rheumatology 2007;47:542–546.

9 Mittermayr R, Hartinger J, Antonic V, Meinl A, Pfeifer S, Stojadinovic A, Schaden W, Redl H: Extracorporeal shock wave therapy (ESWT) minimizes ischemic tissue necrosis irrespective of application time and promotes tissue revascularization by stimulating angiogenesis. Ann Surg 2011;253:1024–1032.

10 Wang CJ, Huang HY, Pai CH: Shock wave-enhanced neovascularization at the tendon-bone junction: an experiment in dogs. J Foot Ankle Surg 2002;41:16–22.

11 Ciampa AR, de Prati AC, Amelio E, Cavalieri E, Persichini T, Colasanti M, Musci G, Marlinghaus E, Suzuki H, Mariotto S: Nitric oxide mediates anti-inflammatory action of extracorporeal shock waves. FEBS Lett 2005;579:6839–6845.

12 Maier M, Schmitz C: Shock wave therapy: what really matters. Ultrasound Med Biol 2008;34:1868–1869.

13 Khan KM, Scott A: Mechanotherapy: how physical therapists' prescription of exercise promotes tissue repair. Br J Sports Med 2009;43:247–252.

14 Abate M, Silbernagel KG, Siljeholm C, Di Iorio A, De Amicis D, Salini V, Werner S, Paganelli R: Pathogenesis of tendinopathies: inflammation or degeneration? Arthritis Res Ther 2009;11:235.

15 Woodwell DA, Cherry DK: National ambulatory medical care survey: 2002 summary. Adv Data 2004: 1–44.

16 Labor USDo: Bureau of labor statistics. http://www-blsgov (accessed August 8, 2005).

17 Furia JP: High-energy extracorporeal shock wave therapy as a treatment for insertional achilles tendinopathy. Am J Sports Med 2016;34:733–740.

18 Lohrer H, Schöll J, Arentz S: Achilles tendinopathy and patellar tendinopathy. Results of radial shockwave therapy in patients with unsuccessfully treated tendinoses. Sportverletzung Sportschaden 2002;16: 108–114.

19 van Leeuwen MT, Zwerver J, van den Akker-Scheek I: Extracorporeal shockwave therapy for patellar tendinopathy: a review of the literature. Br J Sports Med 2009;43:163–168.

20 Bara T, Synder M: Nine-years experience with the use of shock waves for treatment of bone union disturbances. Ortop Traumatol Rehabil 2007;9:254–258.

21 Maffulli N, Longo UG, Denaro V: Novel approaches for the management of tendinopathy. J Bone Joint Surg Am 2010;92:2604–2613.

22 Maffulli N, Khan KM, Puddu G: Overuse tendon conditions: time to change a confusing terminology. Arthroscopy 1998;14:840–843.

23 Sems A, Dimeff R, Iannotti JP: Extracorporeal shock wave therapy in the treatment of chronic tendinopathies. J Am Acad Orthop Surg 2006;14:195–204.

24 van der Worp H, van den Akker-Scheek I, van Schie H, Zwerver J: ESWT for tendinopathy: technology and clinical implications. Knee Surgery Sports Traumatol Arthroscop 2012;21:1451–1458.

25 Hausdorf J, Lemmens MAM, Kaplan S, Marangoz C, Milz S, Odaci E, Korr H, Schmitz C, Maier M: Extracorporeal shockwave application to the distal femur of rabbits diminishes the number of neurons immunoreactive for substance P in dorsal root ganglia l5. Brain Res 2008;1207:96–101.

26 Maier M, Averbeck B, Milz S, Refior HJ, Schmitz C: Substance P and prostaglandin E2 release after shock wave application to the rabbit femur. Clin Orthop Relat Res 2003;406:237–245.

27 Wang CJ, Wang FS, Yang KD, Weng LH, Hsu CC, Huang CS, Yang LC: Shock wave therapy induces neovascularization at the tendon-bone junction. A study in rabbits. J Ortho Res 2003;21:984–989.

28 Vetrano M, d'Alessandro F, Torrisi MR, Ferretti A, Vulpiani MC, Visco V: Extracorporeal shock wave therapy promotes cell proliferation and collagen synthesis of primary cultured human tenocytes. Knee Surg Sports Traumatol Arthroscop 2011;19: 2159–2168.

29 Han SH, Lee JW, Guyton GP, Parks BG, Courneya JP, Schon LC: J.Leonard Goldner award 2008: effect of extracorporeal shock wave therapy on cultured tenocytes. Foot Ankle Int 2009;30:93–98.

30 Bosch G, de Mos M, van Binsbergen R, van Schie HT, van de Lest CH, van Weeren PR: The effect of focused extracorporeal shock wave therapy on collagen matrix and gene expression in normal tendons and ligaments. Equine Vet J 2009;41:335–341.

31 Hsu RW, Hsu WH, Tai CL, Lee KF: Effect of shockwave therapy on patellar tendinopathy in a rabbit model. J Orthop Res 2004;22:221–227.

32 Brañes J, Contreras HR, Cabello P, Antonic V, Guiloff LJ, Brañes M: Shoulder rotator cuff responses to extracorporeal shockwave therapy: morphological and immunohistochemical analysis. Shoulder Elbow 2012;4:163–168.

33 Daecke W, Kusnierczak D, Loew M: Long-term effects of extracorporeal shockwave therapy in chronic calcific tendinitis of the shoulder. J Shoulder Elbow Surg 2002;11:476–480.

34 Galasso O, Amelio E, Riccelli DA, Gasparini G: Short-term outcomes of extracorporeal shock wave therapy for the treatment of chronic non-calcific tendinopathy of the supraspinatus: a double-blind, randomized, placebo-controlled trial. BMC Musculoskelet Dis 2012;13:86.

35 Gerdesmeyer L, Henne M, Göbel M, Diehl P: Physical principles and generation of shockwaves; in Gerdesmeyer L, Scott Weil L (ed): Extracorporeal Shockwave Therapy. Data Trace Publishing Company, 2007, Chapter 2, pp 11–20.

36 Rompe JD, Rumler F, Hopf C, Nafe B, Heine J: Extracorporal shock wave therapy for calcifying tendinitis of the shoulder. Clin Orthop Relat Res 1995; 321:196–201.

37 Rebuzzi E, Coletti N, Schiavetti S, Giusto F: Arthroscopy surgery versus shock wave therapy for chronic calcifying tendinitis of the shoulder. J Orthop Traumatol 2008;9:179–185.

38 Ioppolo F, Rompe JD, Furia JP, Cacchio A: Clinical application of shock wave therapy (SWT) in musculoskeletal disorders. Eur J Phys Rehabil Med 2014;50: 217–230.

39 Vavken P, Holinka J, Rompe JD, Dorotka R: Focused extracorporeal shock wave therapy in calcifying tendinitis of the shoulder: a meta-analysis. Sports Health 2009;1:137–144.

40 Crowther MA, Bannister GC, Huma H, Rooker GD: A prospective, randomised study to compare extracorporeal shock-wave therapy and injection of steroid for the treatment of tennis elbow. J Bone Joint Surg Br 2002;84:678–679.

41 Speed CA, Nichols D, Richards C, Humphreys H, Wies JT, Burnet S, Hazleman BL: Extracorporeal shock wave therapy for lateral epicondylitis – a double blind randomised controlled trial. J Orthop Res 2002;20:895–898.

42 Buchbinder R, Green SE, Youd JM, Assendelft WJ, Barnsley L, Smidt N: Systematic review of the efficacy and safety of shock wave therapy for lateral elbow pain. J Rheumatol 2006;33:1351–1363.

43 Chung B, Wiley JP: Effectiveness of extracorporeal shock wave therapy in the treatment of previously untreated lateral epicondylitis: a randomized controlled trial. Am J Sports Med 2004;32:1660–1667.

44 Leal C. DL, Juschten J: Shockwave Therapy for Patellar Tendinopathy in Patients with Total Knee Arthroplasties. Transactions of the 17th International Congress of the ISMST, 2014.

45 Melikyan EY, Shahin E, Miles J, Bainbridge LC: Extracorporeal shock-wave treatment for tennis elbow. A randomised double-blind study. J Bone Joint Surg Br 2003;85:852–855.

46 Staples MP, Forbes A, Ptasznik R, Gordon J, Buchbinder R: A randomized controlled trial of extracorporeal shock wave therapy for lateral epicondylitis (tennis elbow). J Rheumatol 2008;35:2038–2046.

47 Rompe JD, Decking J, Schoellner C, Theis C: Repetitive low-energy shock wave treatment for chronic lateral epicondylitis in tennis players. Am J Sports Med 2004;32:734–743.

48 Rompe JD, Hope C, Kullmer K, Heine J, Burger R: Analgesic effect of extracorporeal shock-wave therapy on chronic tennis elbow. J Bone Joint Surg Br 1996;78:233–237.

49 Thiele S, Thiele R, Gerdesmeyer L: Lateral epicondylitis: this is still a main indication for extracorporeal shockwave therapy. Int J Surg 2015;24:165–170.

50 Pettrone FA, McCall BR: Extracorporeal shock wave therapy without local anesthesia for chronic lateral epicondylitis. J Bone Joint Surg Am 2005;87:1297–1304.

51 Radwan YA, ElSobhi G, Badawy WS, Reda A, Khalid S: Resistant tennis elbow: shock-wave therapy versus percutaneous tenotomy. Int Orthop 2007;32:671–677.

52 Ozturan KE, Yucel I, Cakici H, Guven M, Sungur I: Autologous blood and corticosteroid injection and extracoporeal shock wave therapy in the treatment of lateral epicondylitis. Orthopedics 2010;33:84–91.

53 Vara F, Garzón N, Ortega G, Alarcon JG, Lopez E: Treatment of the Patellar Tendinitis with Local Application of Extracorporeal Shock Waves. Abstract 41 from the 3rd Congress of the International Society for Medical Shockwave Treatment Naples, 2000.

54 Ogden JA, Alvarez RG, Levitt RL, Johnson JE, Marlow ME: Electrohydraulic high-energy shock-wave treatment for chronic plantar fasciitis. J Bone Joint Surg Am 2004;86:2216–2228.

55 Taunton J, Taunton KM, Khan KM: Treatment of patellar tendinopathy with extracorporeal shock wave therapy. BCMJ 2003;45.

56 Peers KH, Lysens RJ, Brys P, Bellemans J: Cross-sectional outcome analysis of athletes with chronic patellar tendinopathy treated surgically and by extracorporeal shock wave therapy. Clin J Sport Med 2003;13:79–83.

57 Vulpiani MC, Vetrano M, Savoia V, Di Pangrazio E, Trischitta D, Ferretti A: Jumper's knee treatment with extracorporeal shock wave therapy: a long-term follow up observational study. J Sports Med Phys Fitness 2007;47:323–328.

58 Wang CJ, Chen HS, Chen CE, Yang KD: Treatment of nonunions of long bone fractures with shock waves. Clin Orthop Relat Res 2001;387:95–101.

59 Leal C, Lopez JC, Herrera JM, Reyes OE, Cortes M: Shockwave Biosurgery and Autologous Growth Factors Combined Therapy in Severe Patellar Tendinopathies. Transactions of the 9th International Congress of the ISMST, 2006.

60 Furia JP, Rompe JD, Cacchio A, Del Buono A, Maffulli N: A single application of low-energy radial extracorporeal shock wave therapy is effective for the management of chronic patellar tendinopathy. Knee Surg Sports Traumatol Arthrosc 2013;21:346–350.

61 van der Worp H, Zwerver J, Hamstra M, van den Akker-Scheek I, Diercks RL: No difference in effectiveness between focused and radial shockwave therapy for treating patellar tendinopathy: a randomized controlled trial. Knee Surg Sports Traumatol Arthrosc 2014;22:2026–2032.

62 Serrano E, Criado JC: ESWT Therapy in Patellar Tendinopathy Comparison of 2 Protocols. Transactions of the 17th International Congress of the ISMST, 2014.

63 Zwerver J, Hartgens F, Verhagen E, van der Worp H, van den Akker-Scheek I, Diercks RL: No effect of extracorporeal shockwave therapy on patellar tendinopathy in jumping athletes during the competitive season: a randomized clinical trial. Am J Sports Med 2011;39:1191–1199.

64 Vetrano M, Castorina A, Vulpiani MC, Baldini R, Pavan A, Ferretti A: Platelet-rich plasma versus focused shock waves in the treatment of jumper's knee in athletes. Am J Sports Med 2013;41:795–803.

65 Burr DB: Why bones bend but don't break. J Musculoskelet Neuronal Interact 2011;11:270–285.

66 Wang CJ, Wang FS, Yang KD, Huang CC, Lee MS, Chan YS, Wang JW, Ko JY: Treatment of osteonecrosis of the HIP: comparison of extracorporeal shockwave with shockwave and alendronate. Arch Orthop Trauma Surg 2008;128:901–908.

67 Maitra RS, Johnson DL: Stress fractures. Clinical history and physical examination. Clin Sports Med 1997;16:259–274.

68 Maier M, Milz S, Tischer T, Münzing W, Manthey N, Stäbler A, Holzknecht N, Weiler C, Nerlich A, Refior HJ, Schmitz C: Influence of extracorporeal shockwave application on normal bone in an animal model in vivo: scintigraphy, MRI and histopathology. J Bone Joint Surg 2002;84:592–599.

69 Valchanou VD, Michailov P: High energy shock waves in the treatment of delayed and nonunion of fractures. Int Orthop 1991;15:181–184.

70 Xu ZH, Jiang Q, Chen DY, Xiong J, Shi DQ, Yuan T, Zhu XL: Extracorporeal shock wave treatment in nonunions of long bone fractures. Int Orthop 2008;33:789–793.

71 Aicher A, Heeschen C, Sasaki Ki, Urbich C, Zeiher AM, Dimmeler S: Low-energy shock wave for enhancing recruitment of endothelial progenitor cells: a new modality to increase efficacy of cell therapy in chronic hind limb ischemia. Circulation 2006;114:2823–2830.

72 Chen YJ, Wurtz T, Wang CJ, Kuo YR, Yang KD, Huang HC, Wang FS: Recruitment of mesenchymal stem cells and expression of TGF-β1 and vegf in the early stage of shock wave-promoted bone regeneration of segmental defect in rats. J Orthop Res 2004;22:526–534.

73 Martini L, Giavaresi G, Fini M, Borsari V, Torricelli P, Giardino R: Early effects of extracorporeal shock wave treatment on osteoblast-like cells: a comparative study between electromagnetic and electrohydraulic devices. J Traumatol Inj Infect Crit Care 2006;61:1198–1206.

74 Martini L, Giavaresi G, Fini M, Torricelli P, de Pretto M, Schaden W, Giardino R: Effect of extracorporeal shock wave therapy on osteoblastlike cells. Clin Orthop Relat Res 2003;413:269–280.

75 Graff J, Pastor J, Senge T, Richter K-D, Funke W: The effect of high energy shock waves on bony tissue – an experimental study. J Urol 1987;137:278A.

76 Biedermann R, Martin A, Handle G, Auckenthaler T, Bach C, Krismer M: Extracorporeal shock waves in the treatment of nonunions. J Traumatol Inj Infect Crit Care 2003;54:936–942.

77 Birnbaum K, Wirtz D, Siebert C, Heller K: Use of extracorporeal shock-wave therapy (ESWT) in the treatment of non-unions. Arch Orthop Traumatol Surg 2002;122:324–330.

78 Alves EM, Angrisani AT, Santiago MB: The use of extracorporeal shock waves in the treatment of osteonecrosis of the femoral head: a systematic review. Clin Rheumatol 2009;28:1247–1251.

79 Ma HZ, Zeng BF, Li XL: Upregulation of VEGF in subchondral bone of necrotic femoral heads in rabbits with use of extracorporeal shock waves. Calcified Tissue Int 2007;81:124–131.

80 Furia JP, Rompe JD, Cacchio A, Maffulli N: Shock wave therapy as a treatment of nonunions, avascular necrosis, and delayed healing of stress fractures. Foot Ankle Clin 2010;15:651–662.

81 Hotzinger A, Radleb L, Lauber US, Lauber H, Platzekc P, Ludwig J: Mri-Guided SWT of Multiple Stress Fractures of the Tibia. Transactions of the ISMST 2nd International ISMST Congress, London, 1999.

82 Leal C, Herrera JM, Murillo M, Duran R, Reyes OE, Lopez JC: Extracorporeal Shockwave Therapy in Tibial Stress Fractures. Abstracts of the International Society of Arthroscopy Knee surgery and Orthopedic Sports Medicine ISAKOS Biennial Congress, Hollywood USA, 2005.

83 Leal C: Shockwave Biosurgery for Stress Fractures. Transactions of the ISMST 9th International ISMST Congress, Rio de Janeiro, 2006.

84 Herrera JM, LC, Murillo M, Duran R, Lopez JC, Reyes OE: [Treatment of tibial stress fractures in high perfonrmance athletes with extracorporeal shockwave lithotripsy] Revista de la Sociedad Colombiana de Cirugia Ortopedica y Traumatologia SCCOT 2005;19:73–80.

85 Leal C, Herrera JM, Murillo M, Duran R, Reyes OE, Lopez JC: ESWT in High Performance Athletes with Tibial Stress Fractures. Transactions of the ISMST 5th International ISMST Congress, Winterthur, 2002.

Prof. Dr. Carlos Leal, MD
Fenway Medical Shockwave Medicine Center
Postgrado de Ortopedia y Traumatologia Universidad El Bosque
Carrera 7B Bis # 132–38, Office 802
Bogota, DC 110121 (Colombia)
E-Mail chazleal@gmail.com

Wang C-J, Schaden W, Ko J-Y (eds): Shockwave Medicine.
Transl Res Biomed. Basel, Karger, 2018, vol 6, pp 87–101 (DOI: 10.1159/000485064)

Preclinical and Clinical Application of Extracorporeal Shockwave for Ischemic Cardiovascular Disease

Hon-Kan Yip[a, b] · Fan-Yen Lee[c] · Kuan-Hung Chen[d] · Pei-Hsun Sung[a] · Cheuk-Kwan Sun[e]

[a]Division of Cardiology, Department of Internal Medicine, [b]Center for Shockwave Medicine and Tissue Engineering, [c]Division of Thoracic and Cardiovascular Surgery, and [d]Department of Anesthesiology, Kaohsiung Chang Gung Memorial Hospital and Chang Gung University College of Medicine, Kaohsiung, and [e]Department of Emergency Medicine, E-Da Hospital, I-Shou University School of Medicine for International Students, Kaohsiung, Taiwan

Abstract

Despite the availability of state-of-the-art treatment, there are still a significant number of patients with severe and diffuse atherosclerotic-obstructive coronary artery disease who are not only unsuitable for treatment of coronary artery disease, but also poor responders to medications. Additionally, atherosclerotic peripheral artery occlusive disease (PAOD) that affects 8–12% of Americans >65 years of age is associated with a major decline in functional status, increased myocardial infarction, and stroke rates as well as increased risk of ischemic amputation. Current treatment strategies for claudication have limitations. Cell therapy has appeared as a potentlal to improve healing of ischemic heart, repopulate injured myocardium and restore cardiac function as well as for PAOD. The tremendous hope and potential of stem cell therapy is well understood, yet recent trials involving cell therapy for cardiovascular diseases have yielded mixed results, that is, with inconsistent data thereby readdressing controversies and unresolved questions regarding stem cell efficacy for ischemic cardiac disease treatment. These issues raise the need for developing an alternative strategic management with safety and efficacy for those of cardiovascular disease patients who are refractory to conventional therapy. Growing data have demonstrated that low energy of extracorporeal shockwave therapy (ESWT) dramatically improved chronic ischemic heart disease and hind limb ischemia, and preserving neurological function after ischemic stroke mainly through enhancing angiogenesis, upregulating the expression of stromal cell-derived factor-1α, recruiting endothelial progenitor cells/mesenchymal stem cells, as well as suppressing inflammation, generation of oxidative stress and cell apoptosis. Accordingly, this book chapter reviews the current data, that is, experimental studies and clinical trials focusing on the safety and efficacy as well as the underlying mechanisms of ESWT on improving ischemia-related organ dysfunction. © 2018 S. Karger AG, Basel

Extracorporeal shockwave lithotripsy (ESWL) is a well-established treatment option for urolithiasis [1–3]. For almost 30 years, ESWL has been clinically implemented as an effective treatment to disintegrate urinary stones. The underlying mechanism of disintegrated urinary stones is the fragmentation that is similar to the fracture of any brittle object. It represents a process whereby cracks form as a result of stresses generated by applied shockwaves [1, 4]. Cracks begin at sites where shockwave-induced stress exceeds a critical value. Further disintegration occurs as a result of growth and coalescence of these cracks under repetitive loading and unloading [4].

For more than 15 years, an innovation of ESWT other than ESWL has been successfully applied to the treatment of chronic tendinitis, painful heels, epicondylitis of the elbow, and delayed bone-fracture healing as well as wound healing with promising results [5–16]. The underlying mechanisms have been suggested to be mainly through the relief of painful sensations [5–7] and the reduction of inflammatory reactions and oxidative stress [17–24].

Extensive data have demonstrated that ESWT is a novel therapeutic modality and its use in promoting tissue regeneration, improving ischemia-related organ dysfunction and augmenting analgesic effect [7, 10, 14, 25–27]. It becomes clear that ESWT not only can represent a valid therapeutic modality [7, 10, 14, 16, 23, 25] but is also convenient, cost-effective, and has negligible complications by applying appropriate energy; it is, therefore, to be considered to bypass many of the problematic issues that would be associated with surgical interventions in our conventionally clinical practice [8, 9, 11–14, 22, 25–27]. Additionally, in view of it is a safe, repeatable, and noninvasive modality, in many cases, it becomes a first-line therapeutic option, as an alternative to surgery, or in combination with some other treatment options [8, 9, 11–14, 22, 26, 27].

Growing data have demonstrated that SW dramatically improved chronic ischemic heart disease and hind limb ischemia, and preserving neurological function after ischemic stroke (IS) mainly through enhancing angiogenesis, upregulating the expression of stromal cell-derived factor (SDF)-1α, recruiting endothelial progenitor cells (EPCs), as well as suppressing inflammation, generation of oxidative stress and cell apoptosis [26–31]. This chapter aims to provide an overview of shockwave therapy on ischemic cardiovascular disease, based on the current knowledge in SW mechanobiology and it is possible to foresee new, interesting and promising applications in the fields of regenerative medicine.

Original Concept of ESWT on Ischemic Organ Dysfunction

Chronic Inflammation Plays a Crucial Role for Developing Cardiovascular Diseases-Impact of ESWT on Suppressing Inflammatory Reaction
Cardiovascular disease is the leading cause of death worldwide [32–34] as it is well known that a majority of cardiovascular diseases are caused by a chronic inflamma-

tory reaction [35]. A plentiful reported data identifies that the participation of numerous inflammatory mediators in the inflammatory process involving vascular cells, is essential to the development of all stages of atherosclerosis, from initiation, to progression, and to the evolvement of complications [34–38]. Pathological and immunohistochemical staining studies clearly demonstrate the preponderance of inflammatory cells in ruptured plaques in fatal cases of acute coronary syndrome [34, 39, 40]. Thus, these experimental results [41–47] further raise the issue of whether formation of atherosclerotic plaque results from inflammatory response cascade. Therefore, a strategic treatment that inhibits an inflammatory mediator may not adequately prevent vascular cell proliferation after arterial injury.

Low energy of ESWT is characterized by employing a sequence of transient pressure disturbances characterized by high peak pressure (100 MPa), fast rise in pressure (<10 ns), rapid propagation, and short lifecycle (10 μs) produced by an appropriate generator and directed to a specific target area with an energy density in the range of 0.003–0.11 mJ/mm^2 [23, 48, 49]. The ESWT has been successfully applied for the treatment of shoulder tendinitis and plantar fasciitis [5, 6, 9, 21, 23, 50–52]. The underlying mechanism for improving tendinitis and plantar fasciitis has been proposed mainly via the anti-inflammatory effect. Additionally, an experimental study has shown that ESWT effectively attenuated acute interstitial cystitis in rodents [25]. The underlying mechanism involved in suppressing this disease entity has been established mainly through the inhibition of inflammation and the generation of oxidative stress [25]. Furthermore, ESWT has been shown to have capacity of inhibition of vascular wall remodeling (i.e., proliferations of intimal and medial layer) after endothelial denudation [21]. The mechanism involving the vessel wall remodeling has been established mainly due to inflammatory reaction [21]. These aforementioned issues [5, 6, 9, 21, 23, 25] highlight that ESWT has the capacity of inhibiting the inflammatory reaction and the generation of oxidative stress, thereby suggesting that that ESWT would be useful for the treatment of chronic inflammation-caused cardiovascular disease [34–47].

Angiogenesis Play a Crucial Role for Improving the Ischemia-Related Cardiovascular Disease-Impact of ESWT on Augmenting Angiogenesis
It is well known that atherosclerotic arterial obstructive disease causes acute coronary syndrome [35–37], IS [38, 53] and peripheral arterial occlusive disease (PAOD) as well as critical limb ischemia (CLI) [54–57], which ultimately causes pump failure and permanent neurological dysfunction, disability, and dependence in activities of daily life as well as death [58–64]. Numerous studies have shown that ischemia-related organ dysfunction play a key role for poor prognostic outcomes in those patients with acute atherosclerotic arterial obstructive syndrome [59–70].

Intriguingly, studies have previously exhibited that ESWT not only can suppress the inflammatory [17–24] reaction but can also enhance angiogenesis/neovascularization in setting of tissue/organ ischemia [22, 26, 28–31, 36]. Additionally, clinical observational studies have also found that ESWT increased angiogenesis in patients with ischemic heart disease [71–74]. These findings [22, 26, 28–31, 36, 71–74] raise the rationale to believe that ESWT may play a fundamental role in restoring the ischemia-related organ dysfunction.

The Mechanisms of ESWT on Anti-Inflammation – Multiple Signaling Pathways

An association between inflammatory reaction and coronary artery disease and tissue/organ damage has been keenly investigated [21, 23, 25–30, 36–40]. These inflammatory biomarkers, include inflammatory cells, proinflammatory cytokines, and inflammatory protein (e.g., C-reactive protein), which have been identified to mainly distribute in circulation, tissue, or organ levels and to directly participate in tissue/organ damage [75–83]. However, currently, effective treatment on inhibiting the inflammatory reaction that would, at the same time, offer benefit to preserve the damaged tissues/organs is still lacking.

Growing data have shown that ESWT has the capacity of suppressing inflammatory reaction [21, 25, 52, 84–87]. Experimental studies have further identified that numerous proinflammatory biomarkers were markedly suppressed by ESWT at the tissue level [21, 25, 85, 87]. Additionally, these studies have displayed that most of these inflammatory biomarkers represent the innate inflammatory response [21, 25, 85–87]. Surprisingly, while the effect of ESWT on suppressing the inflammatory reaction has been universally accepted, the exactly underlying mechanism for ESWT on inhibiting the inflammatory reaction has not yet to be fully clarified. Some experimental studies and review articles have proposed that the intrinsic property of ESWT therapy on reduction of the inflammation is mainly through a broad-spectrum of regulating the signaling pathways of proinflammatory cytokines, such as tumor necrotic factor-1α, nuclear factor-κB, interleukin (IL)-β, IL-6, and innate inflammatory cells such as leukocytes, macrophages (i.e., increased M2 macrophages and decreased M1 macrophages), IL-18+ cells, and CD40+ cells [21, 22, 25–27, 52, 85, 86, 88]. Other experimental studies have shown that low energy ESW caused a significant inhibition of some M1 marker genes (CD80, COX2, CCL5) in M1 macrophages and a significant synergistic effect for some M2 marker genes (ALOX15, MRC1, CCL18) in M2 macrophages, and this therapy also affected cytokine and chemokine production, inducing in particular a significant increase in IL-10 and reduction in IL-1β production [86], as well as ESWT-modulated inflammation via the TLR3 pathway [89]. In this way, Tepekoylu et al. [90] has also found that ESWT augmented macrophage infiltration and increased polarization toward M2 macrophages in the treated animals. Interestingly, Mariotto et al. [91] has previously established that ESWT increased nitric oxide (NO) levels and the subsequently suppressed nuclear factor-kappaB activation that may account, at least in part, for the clinically beneficial action on tissue inflammation.

The Mechanism of ESWT on Enhancing Angiogenesis –Molecular-Cellular Levels

The phenomenon of ischemia-related organ dysfunction is world widely accepted. Accordingly, to restore the blood flow is crucial for improving the organ dysfunction. Vascular endothelial growth factor (VEGF), one angiogenesis factor has been identified to play the crucial role for angiogenesis by a body of previous studies [31, 92, 93]. Thus, exogenous administration of VEGF to ischemic organs was considered an innovative therapeutic option for improving the ischemic tissue and ischemia-related organ dysfunction. This rationale concept had encouraged the investigators to perform some clinical trials [94, 95] to prove the therapeutic impact of exogenous administration of VEGF on reversing the ischemia-related organ dysfunction. However, these clinical trials [94, 96, 97] were withdrawn due to severe side effects such as the formation of aberrant and leaky vessels. These failed clinical trials [94–99] reappraised the enhancing endogenous angiogenesis factors rather than administration of the recumbent (i.e., exogenous) angiogenesis factors. This may be fundamentally a therapeutic option for restoring the blood flow and improving ischemia-related organ dysfunction.

Investigators are always interested in the therapeutic impact of ESW on enhancing endogenous angiogenesis in ischemic tissues/organs, especially in heart and limb ischemia [26, 28–31, 90, 99–102]. In fact, the underlying mechanism of ESWT on enhancing angiogenesis has been keenly investigated by previous studies [26, 28–31, 90, 99–102]. However, the exactly central-core mechanism of ESWT for augmenting angiogenesis in ischemic regions largely remains uncertain. Some investigators have found that ESWT enhanced angiogenesis through VEGFR2 activation and recycling [31, 100] as well as through activating the toll-like receptor 3 signaling [101]. Other investigators have also shown that ESWT enhanced angiogenesis mainly though up-regulating SDF-1α, which in turn recruited the circulating EPCs to the ischemic tissues/organs for angiogenesis [26, 29] as well as enhanced the differentiation of angiogenesis cells [70, 99] and mobilizing endothelial cells into the ischemic region for angiogenesis [102].

ESWT for Ischemic Heart Disease – Experimental Studies

Based on the reports that ESWT can enhance angiogenesis [26, 28–31, 90, 99–102] and restore the blood flow in ischemic region, ESW has been applied to the treatment of ischemia-induced organ dysfunction. To review the literatures, readers will found that Dr. Nishida et al. [103] is the pioneer to utilize the ESWT for ischemic heart disease in pig in vivo. The results showed that this physical therapy can improve ischemia-induced myocardial dysfunction mainly through enhancing angiogenesis and restoring the blood flow/circulation in ischemic region [103]. Because the results are promising and attractive, thereafter, this treatment is going on quite well. In a mini-pig myocardial ischemia model, Dr. Fu et al. [26] found that ESWT reversed ischemia-related left ventricular (LV) dysfunction and remodeling. They established that the underlying mechanism involved on improving LV function and inhibiting the LV remodeling including (1) enhancing angiogenesis, (2) augmenting the small number of vessels in ischemic region, and (3) suppressing inflammatory response, apoptosis and fibrosis as well as the generation of oxidative stress. Additionally, Dr. Fu et al. [26] and Dr. Ito et al. [104] found that ESWT enhanced endogenous angiogenesis factors and neovascularization, especially that of the VEGF, plays a fundamental role for improving myocardial ischemia in pig in vivo [104]. Not only in experimental myocardial ischemic model and in acute myocardial infarction (AMI) experimental model, ESWT has also been found to be effective on improving the heart function after AMI [28, 30, 103–107]. The above-mentioned experimental data support that ESWT may be an alternative therapeutic option for those patients refractory to traditional therapy.

ESWT for Refractory Angina-Clinical Trials

When we look at the literatures, we can find that application of ESWT for patients with clinical settings of angina pectoris/ischemic heart disease [104] is mainly based on the results of those of animal studies that show that ESWT was not only safe but also efficacious [103, 105–107]. Dr. Khattab et al. [108] and Dr. Ito et al. [109] are the first investigators to apply ESWT on those patients with angina pectoris and refractory to conventional therapy. They demonstrated that ESWT improved symptoms and myocardial ischemia in patients with severe coronary artery disease [109]. Of particular importance was that there were no procedural complications or adverse effects in the ESW-treated patients. Therefore, they concluded that ESWT is an effective and noninvasive treatment for ischemic heart disease [109]. Thereafter, ESWT for patients with chronic angina pectoris/ischemic heart disease has sprouted up worldwide [71–74, 110–115]. These clinical trials have demonstrated that ESWT improvement in symptoms, regional myocardial blood flow, quality of life parameters, and ischemic threshold during exercise in patients with chronic refractory angina pectoris [71–74, 110–115]. Currently, ESWT is one of the accessory therapeutic tools for those chronic angina pectoris patients who are refractory to optimal conventional therapy. Therefore, ESWT appears to be an effective, safe, and noninvasive angiogenic approach in cardiovascular medicine and its indication could be extended to a variety of ischemic diseases in the near future.

ESWT for CLI-Experimental Studies

PAOD is a remarkably highly prevalent arterial atherosclerotic occlusive syndrome that affects approximately 8–12 million individuals in the United States and is associated with significant morbidity and mortality [116]. Importantly, patients with PAOD may develop CLI at a later stage of the disease [116, 117]. Despite the existence of catheter-based and bypass surgical interventions – the two standard methods with the highest successful rate – the procedural failure rate, short-term, and long-term untoward clinical outcomes in some patients remain clinically unresolved issues. Development of a more cost-effective option to treat the patients with CLI who are unsuitable for either surgical or percutaneous intervention, therefore, is obligatory and urgently needed.

A previous experimental study reported by Dr. Yeh et al. [29] has demonstrated that ESWT significantly improved the rodent CLI mainly through enhancing angiogenesis and restoring the blood flow in the CLI region. Additionally, they found that combined treatment with EPCs and ESW is superior to either of these methods alone in improving ischemia in rodent CLI. They have established ESWT-enhanced numbers of CD31+, vWF+, and CXCR4+ cells in the ischemic area through enhancing local production of SDF-1α, suggesting that the enhanced accumulation of EPCs in the ischemic area after ESW therapy may be due to both an increased SDF-1α level in ischemic tissues for EPC homing and the retaining effect of this chemokine on implanted EPCs.

ESWT for CLI-Clinical Application

Another clinical application of ESWT is for those of patients with PAOD [29, 118–124]. Seventeen years ago, Dr. De Sanctis et al. [118] first invested the impact of ESWT on CLI patients. Impressively, they found that ESWT augmented angiogenesis, increased microcirculation, restored and the blood flow and improved the pain-free walking distance [118]. The results of this study [118] encouraged more and more other investigators [119, 125] to further assess the therapeutic modality of ESW on those of PAOD patients. Consistently, these later studies have also shown a significant reduction in the Fontaine stage and significant improvements in pain-free walking distance and maximum walking distance in PAOD patients [119, 120, 125]. Currently, ESWT has been recommended as a potentially novel treatment for intermittent claudication [122].

Innovative Option for Tissue Regeneration – ESWT in Our Future Clinical Practice

Combination of ESW and Stem Cell Therapy-Experimental Study and Clinical Trial
Combination of 2 or more drugs for treatment of one disease, such as hypertension, diabetes mellitus, or acute respiratory distress syndrome, and so on, is a very common strategic management for the patients in our daily clinical practice. In fact, such strategic management is almost always to be identified to get the additional benefit for the patients. Based on this concept, Dr. Sheu et al. [30] utilized a swine AMI model to test

the hypothesis that combined therapy with ESWT and autologous bone marrow-derived mesenchymal stem cells (BMDMSCs) is superior to either therapy alone for alleviating LV dysfunction. The results of their study demonstrated that combined ESW and BMDMSC therapy is superior to either therapy alone for improving LVEF, reducing infarct size, and inhibiting LV remodeling mainly through the enhancement of angiogenesis, suppressions of inflammatory reaction, apoptosis, and generation of oxidative stress [30].

Intriguingly, based on the fact, in post-infarction heart failure patients, intracoronary administration of autologous BMDMSCs was identified an impairment of homing of BMCs to the target area and ESWT increased homing factors in the target tissue, resulting in enhanced retention of applied BMCs, Dr. Assmus et al. [72] performed a clinical trial of combined ESW-BMDMSC for AMI patients with post-infarction heart failure. The results from this clinical trial showed that among patients with post-infarction chronic heart failure, ECSW-facilitated intracoronary administration of BMDMSCs vs. shockwave treatment alone resulted in a significant, albeit modest, mprovement in LV function at 4 months after the treatment. These experimental studies [30] and clinical trials [72] may highlight the therapeutic potential of combined ESW-stem cell for patients with ischemic heart disease who are refractory to conventional therapy.

ESWT for IS-Experimental Study
Stroke continues to be a fundamental disease in developed countries [54, 126]. Not only is it associated with a high mortality/morbidity and prevalence in the elderly, but it also exhibits an increasing incidence to attack in younger cohort, which is the principal manpower in the economic composition of each country [127, 128]. In the background of cerebrovascular disease entities, besides acute IS, cerebrovascular insufficiency and its associated risk of dementia also contribute to the social economic burden [129, 130]. The global increase in the incidence of cerebrovascular diseases and lack of satisfactory management imply the requirement for an effective therapeutic strategy to reduce the mortality and neurologic sequelae as well as permanent disability [131].

Intriguingly, in acute IS by left internal carotid artery occlusion in rats, Dr. Yuen et al. [27] investigated the effect of ESWT on brain-infarction volume and neurological function. By day 28 after acute IS, the brain magnetic resonance imaging demonstrated that when compared with the IS animals, the brain-infarction volume was significantly reduced in IS animals after receiving ESWT. Additionally, by day 21 after acute IS, the sensorimotor-functional test identified a higher frequency of turning movement to left in IS animals than that in IS + ESWT animals, suggesting that the neurological function was improved in IS animals after undergoing the ESWT [27]. Besides, they found that ESWT improved the outcomes after acute IS in rats mainly through augmenting the expressions of angiogenesis and heat-shock protein 70 and by inhibiting the expressions of inflammation, oxidative stress, apoptosis, and brain edema [27]. Currently, there are no data to report the clinical application of ESWT in patients with cerebrovascular disease entity. However, the results of the experimental study may raise the need of a prospective clinical study to assess the therapeutic impact of ECSWT on cerebrovascular disease patients [27].

ESWT for Diabetic Neuropathy – Experimental Study

Diabetic peripheral neuropathy (DPN) is the most common problem of diabetes. Approximately 60% of all diabetic patients have been found to develop DPN, the most common microvascular complication [132, 133]. In fact, DPN, which has been demonstrated as a chronic inflammatory state [87, 134] is characterized by a progressive distal-to-proximal degeneration of peripheral nerves and results in sensory symptoms, including pain, weakness, and/or loss of sensation. DPN is associated with significant morbidity and mortality, and other than tight glycemic control, no effective treatment methods are available.

Interestingly, by using an experimental model study, Dr. Chen et al. [87] tested the hypothesis that ESWT can effectively protect the sciatic nerve from DM-induced neuropathy in leptin-deficient (ob/ob) mice. Their results displayed that ESWT protected the sciatic nerve against DM-induced neuropathy.

References

1 Chaussy C, Brendel W, Schmiedt E: Extracorporeally induced destruction of kidney stones by shock waves. Lancet 1980;2:1265–1268.

2 Fuchs G, Miller K, Rassweiler J, Eisenberger F: Extracorporeal shock-wave lithotripsy: one-year experience with the dornier lithotripter. Eur Urol 1985;11:145–149.

3 Rassweiler JJ, Renner C, Chaussy C, Thuroff S: Treatment of renal stones by extracorporeal shock-wave lithotripsy: an update. Eur Urol 2001;39:187–199.

4 Lokhandwalla M, Sturtevant B: Fracture mechanics model of stone comminution in ESWL and implications for tissue damage. Phys Med Biol 2000;45:1923–1940.

5 Rompe JD, Zoellner J, Nafe B: Shock wave therapy versus conventional surgery in the treatment of calcifying tendinitis of the shoulder. Clin Orthop Relat Res 2001:72–82.

6 Rompe JD, Decking J, Schoellner C, Nafe B: Shock wave application for chronic plantar fasciitis in running athletes. A prospective, randomized, placebo-controlled trial. Am J Sports Med 2003;31:268–275.

7 Wang L, Qin L, Lu HB, Cheung WH, Yang H, Wong WN, Chan KM, Leung KS: Extracorporeal shock wave therapy in treatment of delayed bone-tendon healing. Am J Sports Med 2008;36:340–347.

8 Wang CJ, Chen HS, Chen WS, Chen LM: Treatment of painful heels using extracorporeal shock wave. J Formos Med Assoc 2000;99:580–583.

9 Wang CJ, Ko JY, Chen HS: Treatment of calcifying tendinitis of the shoulder with shock wave therapy. Clin Orthop Relat Res 2001:83–89.

10 Wang CJ, Chen HS, Chen CE, Yang KD: Treatment of nonunions of long bone fractures with shock waves. Clin Orthop Relat Res 2001:387:95–101.

11 Wang CJ, Chen HS: Shock wave therapy for patients with lateral epicondylitis of the elbow: a one- to two-year follow-up study. Am J Sports Med 2002;30:422–425.

12 Wang CJ, Yang KD, Wang FS, Chen HH, Wang JW: Shock wave therapy for calcific tendinitis of the shoulder: a prospective clinical study with two-year follow-up. Am J Sports Med 2003;31:425–430.

13 Wang CJ, Wang FS, Huang CC, Yang KD, Weng LH, Huang HY: Treatment for osteonecrosis of the femoral head: comparison of extracorporeal shock waves with core decompression and bone-grafting. J Bone Joint Surg Am 2005;87:2380–2387.

14 Kuo YR, Wang CT, Wang FS, Chiang YC, Wang CJ: Extracorporeal shock-wave therapy enhanced wound healing via increasing topical blood perfusion and tissue regeneration in a rat model of STZ-induced diabetes. Wound Repair Regen 2009;17:522–530.

15 Yang MY, Chiang YC, Huang YT, Chen CC, Wang FS, Wang CJ, Kuo YR: Serum proteomic analysis of extracorporeal shock wave therapy-enhanced diabetic wound healing in a streptozotocin-induced diabetes model. Plast Reconstr Surg 2014;133:59–68.

16 Russo S, Sadile F, Esposito R, Mosillo G, Aitanti E, Busco G, Wang CJ: Italian experience on use of E.S.W. Therapy for avascular necrosis of femoral head. Int J Surg 2015;24:188–190.

17 Wang CJ, Huang HY, Pai CH: Shock wave-enhanced neovascularization at the tendon-bone junction: an experiment in dogs. J Foot Ankle Surg 2002;41:16–22.

18 Wang CJ: An overview of shock wave therapy in musculoskeletal disorders. Chang Gung Med J 2003;26:220–232.

19 Wang CJ, Wang FS, Yang KD, Weng LH, Hsu CC, Huang CS, Yang LC: Shock wave therapy induces neovascularization at the tendon-bone junction. A study in rabbits. J Orthop Res 2003;21:984–989.

20 Wang CJ, Wang FS, Yang KD, Weng LH, Sun YC, Yang YJ: The effect of shock wave treatment at the tendon-bone interface-an histomorphological and biomechanical study in rabbits. J Orthop Res 2005;23:274–280.

21 Shao PL, Chiu CC, Yuen CM, Chua S, Chang LT, Sheu JJ, Sun CK, Wu CJ, Wang CJ, Yip HK: Shock wave therapy effectively attenuates inflammation in rat carotid artery following endothelial denudation by balloon catheter. Cardiology 2010;115:130–144.

22 Kuo YR, Wu WS, Hsieh YL, Wang FS, Wang CT, Chiang YC, Wang CJ: Extracorporeal shock wave enhanced extended skin flap tissue survival via increase of topical blood perfusion and associated with suppression of tissue pro-inflammation. J Surg Res 2007;143:385–392.

23 Sun CK, Shao PL, Wang CJ, Yip HK: Study of vascular injuries using endothelial denudation model and the therapeutic application of shock wave: a review. Am J Transl Res 2011;3:259–268.

24 Cheng JH, Wang CJ: Biological mechanism of shock-wave in bone. Int J Surg 2015;24:143–146.

25 Chen YT, Yang CC, Sun CK, Chiang HJ, Chen YL, Sung PH, Zhen YY, Huang TH, Chang CL, Chen HH, Chang HW, Yip HK: Extracorporeal shock wave therapy ameliorates cyclophosphamide-induced rat acute interstitial cystitis though inhibiting inflammation and oxidative stress-in vitro and in vivo experiment studies. Am J Transl Res 2014;6:631–648.

26 Fu M, Sun CK, Lin YC, Wang CJ, Wu CJ, Ko SF, Chua S, Sheu JJ, Chiang CH, Shao PL, Leu S, Yip HK: Extracorporeal shock wave therapy reverses ischemia-related left ventricular dysfunction and remodeling: molecular-cellular and functional assessment. PLoS One 2011;6:e24342.

27 Yuen CM, Chung SY, Tsai TH, Sung PH, Huang TH, Chen YL, Chai HT, Zhen YY, Chang MW, Wang CJ, Chang HW, Sun CK, Yip HK: Extracorporeal shock wave effectively attenuates brain infarct volume and improves neurological function in rat after acute ischemic stroke. Am J Transl Res 2015;7:976–994.

28 Sheu JJ, Sun CK, Chang LT, Fang HY, Chung SY, Chua S, Fu M, Lee FY, Kao YH, Ko SF, Wang CJ, Yen CH, Leu S, Yip HK: Shock wave-pretreated bone marrow cells further improve left ventricular function after myocardial infarction in rabbits. Ann Vasc Surg 2010;24:809–821.

29 Yeh KH, Sheu JJ, Lin YC, Sun CK, Chang LT, Kao YH, Yen CH, Shao PL, Tsai TH, Chen YL, Chua S, Leu S, Yip HK: Benefit of combined extracorporeal shock wave and bone marrow-derived endothelial progenitor cells in protection against critical limb ischemia in rats. Crit Care Med 2012;40:169–177.

30 Sheu JJ, Lee FY, Yuen CM, Chen YL, Huang TH, Chua S, Chen YL, Chen CH, Chai HT, Sung PH, Chang HW, Sun CK, Yip HK: Combined therapy with shock wave and autologous bone marrow-derived mesenchymal stem cells alleviates left ventricular dysfunction and remodeling through inhibiting inflammatory stimuli, oxidative stress & enhancing angiogenesis in a swine myocardial infarction model. Int J Cardiol 2015;193:69–83.

31 Huang TH, Sun CK: Shock wave therapy enhances angiogenesis through VEGFR2 activation and recycling. Mol Med 2016;22:1.

32 Murray CJ, Lopez AD: Global mortality, disability, and the contribution of risk factors: global burden of disease study. Lancet 1997;349:1436–1442.

33 Murray CJ, Lopez AD: Mortality by cause for eight regions of the world: global burden of disease study. Lancet 1997;349:1269–1276.

34 Braunwald E: Cardiovascular medicine at the turn of the millennium: triumphs, concerns, and opportunities. N Engl J Med 1997;337:1360–1369.

35 Epstein FH, Ross R: Atherosclerosis – an inflammatory disease. N Engl J Med 1999;340:115–126.

36 van der Wal AC, Becker AE, van der Loos CM, Das PK: Site of intimal rupture or erosion of thrombosed coronary atherosclerotic plaques is characterized by an inflammatory process irrespective of the dominant plaque morphology. Circulation 1994;89:36–44.

37 Yip HK, Wu CJ, Chang HW, Yang CH, Yeh KH, Chua S, Fu M: Levels and values of serum high-sensitivity C-reactive protein within 6 hours after the onset of acute myocardial infarction. Chest 2004;126:1417–1422.

38 Yip HK, Wu CJ, Yang CH, Chang HW, Fang CY, Hung WC, Hang CL: Serial changes in circulating concentrations of soluble CD40 ligand and C-reactive protein in patients with unstable angina undergoing coronary stenting. Circ J 2005;69:890–895.

39 Burke AP, Farb A, Malcom GT, Liang YH, Smialek J, Virmani R: Coronary risk factors and plaque morphology in men with coronary disease who died suddenly. N Engl J Med 1997;336:1276–1282.

40 Burke AP, Tracy RP, Kolodgie F, Malcom GT, Zieske A, Kutys R, Pestaner J, Smialek J, Virmani R: Elevated C-reactive protein values and atherosclerosis in sudden coronary death: association with different pathologies. Circulation 2002;105:2019–2023.

41 Remskar M, Li H, Chyu KY, Shah PK, Cercek B: Absence of CD40 signaling is associated with an increase in intimal thickening after arterial injury. Circ Res 2001;88:390–394.

42 Maffia P, Grassia G, Di Meglio P, Carnuccio R, Berrino L, Garside P, Ianaro A, Ialenti A: Neutralization of interleukin-18 inhibits neointimal formation in a rat model of vascular injury. Circulation 2006;114:430–437.

43 Chadjichristos CE, Matter CM, Roth I, Sutter E, Pelli G, Luscher TF, Chanson M, Kwak BR: Reduced connexin43 expression limits neointima formation after balloon distension injury in hypercholesterolemic mice. Circulation 2006;113:2835–2843.

44 Selzman CH, Meldrum DR, Cain BS, Meng X, Shames BD, Ao L, Harken AH: Interleukin-10 inhibits postinjury tumor necrosis factor-mediated human vascular smooth muscle proliferation. J Surg Res 1998;80:352–356.

45 Honda HM, Leitinger N, Frankel M, Goldhaber JI, Natarajan R, Nadler JL, Weiss JN, Berliner JA: Induction of monocyte binding to endothelial cells by mm-ldl: Role of lipoxygenase metabolites. Arterioscler Thromb Vasc Biol 1999;19:680–686.

46 De Caterina R, Libby P, Peng HB, Thannickal VJ, Rajavashisth TB, Gimbrone MA, Jr, Shin WS, Liao JK: Nitric oxide decreases cytokine-induced endothelial activation. Nitric oxide selectively reduces endothelial expression of adhesion molecules and proinflammatory cytokines. J Clin Invest 1995;96:60–68.

47 Tedgui A, Mallat Z: Anti-inflammatory mechanisms in the vascular wall. Circ Res 2001;88:877–887.

48 Ogden JA, Toth-Kischkat A, Schultheiss R: Principles of shock wave therapy. Clin Orthop Relat Res 2001:8–17.

49 Ciampa AR, de Prati AC, Amelio E, Cavalieri E, Persichini T, Colasanti M, Musci G, Marlinghaus E, Suzuki H, Mariotto S: Nitric oxide mediates anti-inflammatory action of extracorporeal shock waves. FEBS Lett 2005;579:6839–6845.

50 Eslamian F, Shakouri SK, Jahanjoo F, Hajialiloo M, Notghi F: Extra corporeal shock wave therapy versus local corticosteroid injection in the treatment of chronic plantar fasciitis, a single blinded randomized clinical trial. Pain Medicine 2016;17:1722–1731.

51 Malliaropoulos N, Crate G, Meke M, Korakakis V, Nauck T, Lohrer H, Padhiar N: Success and recurrence rate after radial extracorporeal shock wave therapy for plantar fasciopathy: a retrospective study. Biomed Res Int 2016;2016:1–8.

52 Sun J, Gao F, Wang Y, Sun W, Jiang B, Li Z: Extracorporeal shock wave therapy is effective in treating chronic plantar fasciitis. Medicine 2017;96:e6621.

53 Chen ZM, Sandercock P, Pan HC, Counsell C, Collins R, Liu LS, Xie JX, Warlow C, Peto R: Indications for early aspirin use in acute ischemic stroke: a combined analysis of 40,000 randomized patients from the Chinese acute stroke trial and the international stroke trial. On behalf of the cast and ist collaborative groups. Stroke 2000;31:1240–1249.

54 Hankey GJ: Stroke: How large a public health problem, and how can the neurologist help? Arch Neurol 1999;56:748–754.

55 Wood AJ, Hiatt WR: Medical treatment of peripheral arterial disease and claudication. N Engl J Med 2001;344:1608–1621.

56 Newman AB, Shemanski L, Manolio TA, Cushman M, Mittelmark M, Polak JF, Powe NR, Siscovick D: Ankle-arm index as a predictor of cardiovascular disease and mortality in the cardiovascular health study. The cardiovascular health study group. Arterioscler Thromb Vasc Biol 1999;19:538–545.

57 Shammas NW: Epidemiology, classification, and modifiable risk factors of peripheral arterial disease. Vasc Health Risk Manag 2007;3:229–234.

58 Nehler MR, Hiatt WR, Taylor LM Jr: Is revascularization and limb salvage always the best treatment for critical limb ischemia? J Vasc Surg 2003;37:704–708.

59 Norgren L, Hiatt WR, Dormandy JA, Nehler MR, Harris KA, Fowkes FG: Inter-society consensus for the management of peripheral arterial disease (TASC II). J Vasc Surg 2007;45(suppl):S5–S67.

60 Erlebacher JA, Weiss JL, Weisfeldt ML, Bulkley BH: Early dilation of the infarcted segment in acute transmural myocardial infarction: role of infarct expansion in acute left ventricular enlargement. J Am Coll Cardiol 1984;4:201–208.

61 Pfeffer MA, Pfeffer JM: Ventricular enlargement and reduced survival after myocardial infarction. Circulation 1987;75:IV93–IV97.

62 Pfeffer MA, Braunwald E, Moyé LA, Basta L, Brown EJ, Cuddy TE, Davis BR, Geltman EM, Goldman S, Flaker GC, Klein M, Lamas GA, Packer M, Rouleau J, Rouleau JL, Rutherford J, Wertheimer JH, Hawkins CM: Effect of captopril on mortality and morbidity in patients with left ventricular dysfunction after myocardial infarction. N Engl J Med 1992;327:669–677.

63 Cohn JN, Ferrari R, Sharpe N: Cardiac remodeling – concepts and clinical implications: a consensus paper from an international forum on cardiac remodeling. Behalf of an international forum on cardiac remodeling. J Am Coll Cardiol 2000;35:569–582.

64 Pfeffer MA, Lamas GA, Vaughan DE, Parisi AF, Braunwald E: Effect of captopril on progressive ventricular dilatation after anterior myocardial infarction. N Engl J Med 1988;319:80–86.

65 Mann DL: Mechanisms and models in heart failure: a combinatorial approach. Circulation 1999;100:999–1008.

66 Yip HK, Chang LT, Chang WN, Lu CH, Liou CW, Lan MY, Liu JS, Youssef AA, Chang HW: Level and value of circulating endothelial progenitor cells in patients after acute ischemic stroke. Stroke 2008;39:69–74.

67 Yip HK, Wu CJ, Fu M, Yeh KH, Yu TH, Hung WC, Chen MC: Clinical features and outcome of patients with direct percutaneous coronary intervention for acute myocardial infarction resulting from left circumflex artery occlusion. Chest 2002;122:2068–2074.

68 Yip HK, Wu CJ, Chang HW, Chen MC, Hang CL, Fang CY, Hsieh YK, Yang CH, Yeh KH, Fu M: Comparison of impact of primary percutaneous transluminal coronary angioplasty and primary stenting on short-term mortality in patients with cardiogenic shock and evaluation of prognostic determinants. Am J Cardiol 2001;87:1184–1188;A4.

69 Sheu JJ, Tsai TH, Lee FY, Fang HY, Sun CK, Leu S, Yang CH, Chen SM, Hang CL, Hsieh YK, Chen CJ, Wu CJ, Yip HK: Early extracorporeal membrane oxygenator-assisted primary percutaneous coronary intervention improved 30-day clinical outcomes in patients with ST-segment elevation myocardial infarction complicated with profound cardiogenic shock. Crit Care Med 2010;38:1810–1817.

70 Yip HK, Chang LT, Sun CK, Youssef AA, Sheu JJ, Wang CJ: Shock wave therapy applied to rat bone marrow-derived mononuclear cells enhances formation of cells stained positive for CD31 and vascular endothelial growth factor. Circ J 2008;72:150–156.

71 Alunni G, Marra S, Meynet I, D'Amico M, Elisa P, Fanelli A, Molinaro S, Garrone P, Deberardinis A, Campana M, Lerman A: The beneficial effect of extracorporeal shockwave myocardial revascularization in patients with refractory angina. Cardiovasc Revasc Med 2015;16:6–11.

72 Assmus B, Walter DH, Seeger FH, Leistner DM, Steiner J, Ziegler I, Lutz A, Khaled W, Klotsche J, Tonn T, Dimmeler S, Zeiher AM: Effect of shock wave-facilitated intracoronary cell therapy on LVEF in patients with chronic heart failure: The CELL-WAVE randomized clinical trial. JAMA 2013;309:1622–1631.

73 Prasad M, Wan Ahmad WA, Sukmawan R, Magsombol EB, Cassar A, Vinshtok Y, Ismail MD, Mahmood Zuhdi AS, Locnen SA, Jimenez R, Callleja H, Lerman A: Extracorporeal shockwave myocardial therapy is efficacious in improving symptoms in patients with refractory angina pectoris – a multicenter study. Coron Artery Dis 2015;26:194–200.

74 Zuoziene G, Laucevicius A, Leibowitz D: Extracorporeal shockwave myocardial revascularization improves clinical symptoms and left ventricular function in patients with refractory angina. Coron Artery Dis 2012;23:62–67.

75 De Scheerder I, Vandekerckhove J, Robbrecht J, Algoed L, De Buyzere M, De Langhe J, De Schrijver G, Clement D: Post-cardiac injury syndrome and an increased humoral immune response against the major contractile proteins (actin and myosin). Am J Cardiol 1985;56:631–633.

76 Epstein FH, Lange LG, Schreiner GF: Immune mechanisms of cardiac disease. N Engl J Med 1994;330:1129–1135.

77 Varda-Bloom N, Leor J, Ohad DG, Hasin Y, Amar M, Fixler R, Battler A, Eldar M, Hasin D: Cytotoxic T lymphocytes are activated following myocardial infarction and can recognize and kill healthy myocytes in vitro. J Mol Cell Cardiol 2000;32:2141–2149.

78 Chien KR: Stress pathways and heart failure. Cell 1999;98:555–558.

79 Chua S, Leu S, Sheu JJ, Lin YC, Chang LT, Kao YH, Yen CH, Tsai TH, Chen YL, Chang HW, Sun CK, Yip HK: Intra-coronary administration of tacrolimus markedly attenuates infarct size and preserves heart function in porcine myocardial infarction. J Inflamm (Lond) 2012;9:21.

80 Sheu JJ, Sung PH, Leu S, Chai HT, Zhen YY, Chen YC, Chua S, Chen YL, Tsai TH, Lee FY, Chang HW, Ko SF, Yip HK: Innate immune response after acute myocardial infarction and pharmacomodulatory action of tacrolimus in reducing infarct size and preserving myocardial integrity. J Biomed Sci 2013;20:82.

81 Yang CH, Sheu JJ, Tsai TH, Chua S, Chang LT, Chang HW, Lee FY, Chen YL, Chung SY, Sun CK, Leu S, Yen CH, Yip HK: Effect of tacrolimus on myocardial infarction is associated with inflammation, ROS, MAP kinase and akt pathways in mini-pigs. J Atheroscler Thromb 2013;20:9–22.

82 Ko SF, Yip HK, Leu S, Lee CC, Sheu JJ, Lee CC, Ng SH, Huang CC, Chen MC, Sun CK: Therapeutic potential of tacrolimus on acute myocardial infarction in minipigs: analysis with serial cardiac magnetic resonance and changes at histological and protein levels. BioMed Res Int 2014;2014:524078.

83 Frangogiannis NG, Smith CW, Entman ML: The inflammatory response in myocardial infarction. Cardiovasc Res 2002;53:31–47.

84 Krukowska J, Wrona J, Sienkiewicz M, Czernicki J: A comparative analysis of analgesic efficacy of ultrasound and shock wave therapy in the treatment of patients with inflammation of the attachment of the plantar fascia in the course of calcaneal spurs. Arch Orthop Trauma Surg 2016;136:1289–1296.

85 Wang HJ, Lee WC, Tyagi P, Huang CC, Chuang YC: Effects of low energy shock wave therapy on inflammatory moleculars, bladder pain, and bladder function in a rat cystitis model. Neurourol Urodyn 2016; 36:1440–1447.

86 Sukubo NG, Tibalt E, Respizzi S, Locati M, d'Agostino MC: Effect of shock waves on macrophages: a possible role in tissue regeneration and remodeling. Int J Surg 2015;24:124–130.

87 Chen YL, Chen KH, Yin TC, Huang TH, Yuen CM, Chung SY, Sung PH, Tong MS, Chen CH, Chang HW, Lin KC, Ko SF, Yip HK: Extracorporeal shock wave therapy effectively prevented diabetic neuropathy. Am J Transl Res 2015;7:2543–2560.

88 Davis TA, Stojadinovic A, Anam K, Amare M, Naik S, Peoples GE, Tadaki D, Elster EA: Extracorporeal shock wave therapy suppresses the early proinflammatory immune response to a severe cutaneous burn injury. Int Wound J 2009;6:11–21.

89 Holfeld J, Tepekoylu C, Kozaryn R, Urbschat A, Zacharowski K, Grimm M, Paulus P: Shockwave therapy differentially stimulates endothelial cells: implications on the control of inflammation via toll-like receptor 3. Inflammation 2014;37:65–70.

90 Tepekoylu C, Lobenwein D, Blunder S, Kozaryn R, Dietl M, Ritschl P, Pechriggl EJ, Blumer MJ, Bitsche M, Schistek R, Kotsch K, Fritsch H, Grimm M, Holfeld J: Alteration of inflammatory response by shock wave therapy leads to reduced calcification of decellularized aortic xenografts in mice†. Eur J Cardiothorac Surg 2015;47:e80–e90.

91 Mariotto S, de Prati AC, Cavalieri E, Amelio E, Marlinghaus E, Suzuki H: Extracorporeal shock wave therapy in inflammatory diseases: molecular mechanism that triggers anti-inflammatory action. Curr Med Chem 2009;16:2366–2372.

92 Jopling HM, Odell AF, Pellet-Many C, Latham AM, Frankel P, Sivaprasadarao A, Walker JH, Zachary IC, Ponnambalam S: Endosome-to-plasma membrane recycling of vegfr2 receptor tyrosine kinase regulates endothelial function and blood vessel formation. Cells 2014;3:363–385.

93 Ganta VC, Choi M, Kutateladze A, Annex BH: Vegf165b modulates endothelial vegfr1-stat3 signaling pathway and angiogenesis in human and experimental peripheral arterial disease. Circ Res 2017;120: 282–295.

94 Henry TD, Annex BH, McKendall GR, Azrin MA, Lopez JJ, Giordano FJ, Shah PK, Willerson JT, Benza RL, Berman DS, Gibson CM, Bajamonde A, Rundle AC, Fine J, McCluskey ER; VIVA Investigators: The viva trial: Vascular endothelial growth factor in ischemia for vascular angiogenesis. Circulation 2003;107:1359–1365.

95 Henry TD, Rocha-Singh K, Isner JM, Kereiakes DJ, Giordano FJ, Simons M, Losordo DW, Hendel RC, Bonow RO, Eppler SM, Zioncheck TF, Holmgren EB, McCluskey ER: Intracoronary administration of recombinant human vascular endothelial growth factor to patients with coronary artery disease. Am Heart J 2001;142:872–880.

96 Epstein SE, Kornowski R, Fuchs S, Dvorak HF: Angiogenesis therapy: amidst the hype, the neglected potential for serious side effects. Circulation 2001; 104:115–119.

97 Carmeliet P: VEGF as a key mediator of angiogenesis in cancer. Oncology 2005;69(suppl 3):4–10.

98 Lee RJ, Springer ML, Blanco-Bose WE, Shaw R, Ursell PC, Blau HM: VEGF gene delivery to myocardium: deleterious effects of unregulated expression. Circulation 2000;102:898–901.

99 Aicher A, Heeschen C, Sasaki K, Urbich C, Zeiher AM, Dimmeler S: Low-energy shock wave for enhancing recruitment of endothelial progenitor cells: a new modality to increase efficacy of cell therapy in chronic hind limb ischemia. Circulation 2006;114:2823–2830.

100 Holfeld J, Tepekoylu C, Blunder S, Lobenwein D, Kirchmair E, Dietl M, Kozaryn R, Lener D, Theurl M, Paulus P, Kirchmair R, Grimm M: Low energy shock wave therapy induces angiogenesis in acute hind-limb ischemia via VEGF receptor 2 phosphorylation. PLoS One 2014;9:e103982.

101 Holfeld J, Tepekoylu C, Reissig C, Lobenwein D, Scheller B, Kirchmair E, Kozaryn R, Albrecht-Schgoer K, Krapf C, Zins K, Urbschat A, Zacharowski K, Grimm M, Kirchmair R, Paulus P: Toll-like receptor 3 signalling mediates angiogenic response upon shock wave treatment of ischaemic muscle. Cardiovasc Res 2016;109:331–343.

102 Tepekoylu C, Wang FS, Kozaryn R, Albrecht-Schgoer K, Theurl M, Schaden W, Ke HJ, Yang Y, Kirchmair R, Grimm M, Wang CJ, Holfeld J: Shock wave treatment induces angiogenesis and mobilizes endogenous cd31/cd34-positive endothelial cells in a hindlimb ischemia model: implications for angiogenesis and vasculogenesis. J Thorac Cardiovasc Surg 2013;146:971–978.

103 Nishida T, Shimokawa H, Oi K, Tatewaki H, Uwatoku T, Abe K, Matsumoto Y, Kajihara N, Eto M, Matsuda T, Yasui H, Takeshita A, Sunagawa K: Extracorporeal cardiac shock wave therapy markedly ameliorates ischemia-induced myocardial dysfunction in pigs in vivo. Circulation 2004;110:3055–3061.

104 Ito K, Fukumoto Y, Shimokawa H: Extracorporeal shock wave therapy for ischemic cardiovascular disorders. Am J Cardiovasc Drugs 2011;11:295–302.

105 Ito K, Fukumoto Y, Shimokawa H: Extracorporeal shock wave therapy as a new and non-invasive angiogenic strategy. Tohoku J Exp Med 2009;219:1–9.

106 Ito Y, Ito K, Shiroto T, Tsuburaya R, Yi GJ, Takeda M, Fukumoto Y, Yasuda S, Shimokawa H: Cardiac shock wave therapy ameliorates left ventricular remodeling after myocardial ischemia-reperfusion injury in pigs in vivo. Coron Artery Dis 2010;21:304–311.

107 Uwatoku T, Ito K, Abe K, Oi K, Hizume T, Sunagawa K, Shimokawa H: Extracorporeal cardiac shock wave therapy improves left ventricular remodeling after acute myocardial infarction in pigs. Coron Artery Dis 2007;18:397–404.

108 Khattab AA, Brodersen B, Schuermann-Kuchenbrandt D, Beurich H, Tolg R, Geist V, Schafer T, Richardt G: Extracorporeal cardiac shock wave therapy: First experience in the everyday practice for treatment of chronic refractory angina pectoris. Int J Cardiol 2007;121:84–85.

109 Ito K, Shimokawa H: [Extracorporeal cardiac shock wave therapy for angina pectoris]. Nihon Rinsho 2008;66:2019–2026.

110 Prinz C, Lindner O, Bitter T, Hering D, Burchert W, Horstkotte D, Faber L: Extracorporeal cardiac shock wave therapy ameliorates clinical symptoms and improves regional myocardial blood flow in a patient with severe coronary artery disease and refractory angina. Case Rep Med 2009;2009:639594.

111 Kikuchi Y, Ito K, Ito Y, Shiroto T, Tsuburaya R, Aizawa K, Hao K, Fukumoto Y, Takahashi J, Takeda M, Nakayama M, Yasuda S, Kuriyama S, Tsuji I, Shimokawa H: Double-blind and placebo-controlled study of the effectiveness and safety of extracorporeal cardiac shock wave therapy for severe angina pectoris. Circ J 2010;74:589–591.

112 Wang Y, Guo T, Cai HY, Ma TK, Tao SM, Chen MQ, Gu Y, Pan JH, Xiao JM, Zhao L, Yang XY, Yang C: [Extracorporeal cardiac shock wave therapy for treatment of coronary artery disease]. Zhonghua Xin Xue Guan Bing Za Zhi 2010;38:711–715.

113 Wang Y, Guo T, Cai HY, Ma TK, Tao SM, Sun S, Chen MQ, Gu Y, Pang JH, Xiao JM, Yang XY, Yang C: Cardiac shock wave therapy reduces angina and improves myocardial function in patients with refractory coronary artery disease. Clin Cardiol 2010;33:693–699.

114 Wang Y, Guo T, Ma TK, Cai HY, Tao SM, Peng YZ, Yang P, Chen MQ, Gu Y: A modified regimen of extracorporeal cardiac shock wave therapy for treatment of coronary artery disease. Cardiovasc Ultrasound 2012;10:35.

115 Cassar A, Prasad M, Rodriguez-Porcel M, Reeder GS, Karia D, DeMaria AN, Lerman A: Safety and efficacy of extracorporeal shock wave myocardial revascularization therapy for refractory angina pectoris. Mayo Clin Proc 2014;89:346–354.

116 Leeper NJ, Kullo IJ, Cooke JP: Genetics of peripheral artery disease. Circulation 2012;125:3220–3228.

117 Gottsater A: Managing risk factors for atherosclerosis in critical limb ischaemia. Eur J Vasc Endovasc Surg 2006;32:478–483.

118 De Sanctis MT, Belcaro G, Nicolaides AN, Cesarone MR, Incandela L, Marlinghaus E, Griffin M, Capodanno S, Ciccarelli R: Effects of shock waves on the microcirculation in critical limb ischemia (CLI) (8-week study). Angiology 2000;51:S69–S78.

119 Belcaro G, Cesarone MR, Dugall M, Di Renzo A, Errichi BM, Cacchio M, Ricci A, Stuard S, Ippolito E, Fano F, Theng A, Kasai M, Hakim G, Acerbi G: Effects of shock waves on microcirculation, perfusion, and pain management in critical limb ischemia. Angiology 2005;56:403–407.

120 Serizawa F, Ito K, Kawamura K, Tsuchida K, Hamada Y, Zukeran T, Shimizu T, Akamatsu D, Hashimoto M, Goto H, Watanabe T, Sato A, Shimokawa H, Satomi S: Extracorporeal shock wave therapy improves the walking ability of patients with peripheral artery disease and intermittent claudication. Circ J 2012;76:1486–1493.

121 Tara S, Miyamoto M, Takagi G, Kirinoki-Ichikawa S, Tezuka A, Hada T, Takagi I: Low-energy extracorporeal shock wave therapy improves microcirculation blood flow of ischemic limbs in patients with peripheral arterial disease: pilot study. J Nippon Med Sch 2014;81:19–27.

122 Cayton T, Harwood A, Smith GE, Chetter I: A systematic review of extracorporeal shockwave therapy as a novel treatment for intermittent claudication. Ann Vasc Surg 2016;35:226–233.

123 Hirsch AT, Criqui MH, Treat-Jacobson D, Regensteiner JG, Creager MA, Olin JW, Krook SH, Hunninghake DB, Comerota AJ, Walsh ME, McDermott MM, Hiatt WR: Peripheral arterial disease detection, awareness, and treatment in primary care. JAMA 2001;286:1317–1324.

124 Ouriel K: Peripheral arterial disease. Lancet 2001;358:1257–1264.

125 Ciccone MM, Notarnicola A, Scicchitano P, Sassara M, Carbonara S, Maiorano M, Moretti B: Shockwave therapy in patients with peripheral artery disease. Adv Ther 2012;29:698–707.

126 Organization WH: The world health report: 1999: Making a difference: message from the Director-General, 1999.

127 Truelsen T, Piechowski-Jozwiak B, Bonita R, Mathers C, Bogousslavsky J, Boysen G: Stroke incidence and prevalence in europe: a review of available data. Eur J Neurol 2006;13:581–598.

128 Johnston SC, Mendis S, Mathers CD: Global variation in stroke burden and mortality: Estimates from monitoring, surveillance, and modelling. Lancet Neurol 2009;8:345–354.

129 Rockwood K, Davis H, MacKnight C, Vandorpe R, Gauthier S, Guzman A, Montgomery P, Black S, Hogan DB, Kertesz A, Bouchard R, Feldman H: The consortium to investigate vascular impairment of cognition: Methods and first findings. Can J Neurol Sci 2003;30:237–243.

130 Roman GC: Stroke, cognitive decline and vascular dementia: the silent epidemic of the 21st century. Neuroepidemiology 2003;22:161–164.

131 Launer LJ, Hofman A: Frequency and impact of neurologic diseases in the elderly of europe: a collaborative study of population-based cohorts. Neurology 2000;54(11 suppl 5):S1–S8.

132 DIABETES F: National Diabetes Fact Sheet, 2011 – Diabetes in Control, 2011.

133 Vincent AM, Callaghan BC, Smith AL, Feldman EL: Diabetic neuropathy: cellular mechanisms as therapeutic targets. Nat Rev Neurol 2011;7:573–583.

134 Liu S, Zheng H, Zhu X, Mao F, Zhang S, Shi H, Li Y, Lu B: Neutrophil-to-lymphocyte ratio is associated with diabetic peripheral neuropathy in type 2 diabetes patients. Diabetes Res Clin Pract 2017;130:90–97.

Hon-Kan Yip, MD
Division of Division of Cardiology, Kaohsiung Chang Gung Memorial Hospital
123, Dapi Road, Niaosung District
Kaohsiung 83301 (Taiwan)
E-Mail han.gung@msa.hinet.net

Wang C-J, Schaden W, Ko J-Y (eds): Shockwave Medicine.
Transl Res Biomed. Basel, Karger, 2018, vol 6, pp 102–108 (DOI: 10.1159/000485065)

Mechanisms Underlying Extracorporeal Shockwave Treatment for Ischemic Cardiovascular Disease

Cheuk-Kwan Sun[a] · Hon-Kan Yip[b, c]

[a]Department of Emergency Medicine, E-Da Hospital, I-Shou University School of Medicine for International Students, Kaohsiung, and [b]Division of Cardiology, Department of Internal Medicine, and [c]Center for Shockwave Medicine and Tissue Engineering, Kaohsiung Chang Gung Memorial Hospital and Chang Gung University College of Medicine, Kaohsiung, Taiwan

Abstract

The role of shockwave (SW) in medicine has evolved from one of being destructive to being constructive. Although low-energy SW has long been shown to be both pro-angiogenic and anti-inflammatory in the treatment of ischemic cardiovascular diseases, not until recently did the underlying molecular mechanisms become tangible. In addition to triggering the release of nitric oxide (NO) and upregulating the expression of vascular endothelial growth factor (VEGF), low-energy SW has also been found to cause the release of various pro-angiogenic and anti-inflammatory molecules. In addition, recent studies have unveiled the role of SW-induced mechanotransduction in the promotion of angiogenesis and cell proliferation. Low-energy SW has also been demonstrated to boost circulating endothelial progenitor cells and their homing to ischemic sites. Clinically, it seems that combining cell treatment and low-energy SW may be a promising therapeutic option for patients with ischemic cardiovascular diseases.   © 2018 S. Karger AG, Basel

Historical Background

Shockwave (SW) is a longitudinal acoustic wave that can propagate inside soft tissue. It is delivered as a single pulse with a duration of around one micro-second with a peak pressure of up to 100 MPa [1]. The role of SW in medicine has been constantly evolving. It was first considered to be a cause of injury to the human body in the 1950s when the world was recovering from the Second World War. Studies on SW in that era focused on the degree of injuries that SW may produce [2, 3]. Almost three decades later, Chaussy et al. [4] first reported that the energy produced by SW could be successfully harnessed for treating patients with extracorporeal renal lithotripsy. Then two decades passed till after the new millennium when clinicians began to realize that low-energy SW (0.03–0.11 mJ/mm^2) can produce subtle biological effects that are of therapeutic potential particularly in the improvement of tissue perfusion. It has been observed that ischemic organs including the heart [5], limb [6], and musculo-skeletal disorders [7, 8] could all be benefited from SW therapy. Following successful clinical application of low-energy SW in the treatment of patients with ischemic heart disease [9–12], the mechanisms underlying the experimental and clinical observations have become an intriguing topic for researchers in both biological and medical fields.

Overview of Biological Effects of Low-Energy Shockwave on the Cardiovascular System

Based on the observed pro-angiogenic and anti-inflammatory actions of low-energy SW, various studies have been conducted at organ, cellular, and molecular levels. Increasing in vivo [6] and in vitro [13, 14] evidence has shown that the anti-inflammatory and angiogenic properties of SW are attributable to the upregulation of endothelial NO synthase (eNOS) [6, 14], vascular endothelial growth factor (VEGF) [5, 6, 15], fibromyalgia syndrome-related tyrosine kinase 1, placental growth factor as well as reduction in brain natriuretic peptide levels [16], thereby suppressing left ventricular remodeling and preserving cardiac function after acute myocardial infarction [17].

SW has also been found to have positive therapeutic impact on vascular diseases. We have previously shown that SW treatment can suppress neointimal proliferation and smooth muscle proliferation in balloon-induced carotid artery injury [18]. Perfusion to ischemic limbs has also been shown to be improved in a rodent model after SW treatment [15]. Enhanced angiogenesis and tissue perfusion in skin grafts and flaps associated with upregulation of eNOS and VEGF expression have also been demonstrated after SW application [19]. The effect of SW on enhancement of tissue perfusion is 2-folded, including the induction of vasodilatation at early postoperative stage and neovascularization at late stage [6].

At genetic level, a study on full-thickness skin isografts in a murine experimental model demonstrated enhanced expressions of CD31 and angiogenesis pathway-specific genes, including cytokines (interleukin-1 beta, interleukin-6, granulocyte-colony stimulating factor, VEGF-A), MMPs (MMP3, MMP9, MMP13), CC chemokines (CCL2, CCL3, CCL4), ELR-CXC chemokines (CXCL1, CXCL2, CXCL5), hypoxia-inducible factors (HIF-1 alpha), and vascular remodeling kinase (Mst1) from 6 h to 7 days after operation. These findings, therefore, underscore the early pro-angiogenic and anti-inflammatory roles of SW in improving post-operative graft perfusion [20].

Recently, the angiogenic effect of SW has been found to be related to cell surface molecules such as VEGFR2 and β1-integrin and their downstream mechanisms through "mechanotransduction" [21, 22]. Through the same process, low-energy SW has also been found to enhance mesenchymal cell proliferation through the mTORC1-FAK signaling pathway [23]. Besides, it has previously been shown that not only can

SW transform bone marrow-derived mononuclear cells into endothelial progenitor cells (EPCs) [24], but it can also recruit circulating EPCs in SW-preconditioned ischemic tissue through the upregulation of the expression of chemoattractant factors [25]. These mechanisms underlying SW treatment for ischemic cardiovascular disease are discussed in details below.

Mechanisms Underlying Low-Energy SW-Induced Biological Activities

Nitric Oxide Production and Anti-Inflammatory Effects
SW-induced "bubble cavitation" effect was first reported a decade ago. Similar to shear stress [26], SW induces localized stress on cell membrane [27] that may partly account for the observed upregulation in eNOS expression. On the other hand, SW has been found to trigger non-enzymatic nitric oxide (NO) production in the presence of hydrogen peroxide and physiological levels of L-arginine [28] (Fig. 1). In addition to the vasodilatatory effect of NO that contributes to the improvement of organ perfusion after SW treatment [15], the anti-inflammatory action of SW-elicited generation of NO has been recognized [1, 13, 14, 18]. Regarding SW-induced anti-inflammatory effects, it has been shown that the expressions of tumor necrosis factor alpha [29], kappa B activation [1], and other acute pro-inflammatory cytokines, CC and CXC chemokines as well as polymorphonuclear neutrophil and macrophage infiltration and extracellular matrix proteolytic activity are all suppressed after SW treatment [18, 30].

VEGF and Angiogenesis
Low-energy SW has been shown to effectively increase VEGF expression in cultured human umbilical vein endothelial cells [5]. Besides, previous in vivo experimental studies also demonstrated an elevated expression of VEGF in ischemic organs following SW treatment [15, 31] (Fig. 1). In terms of mechanisms, SW-induced activation, phosphorylation, and recycling of VEGF receptor 2 (VEGFR2) have been found to contribute to angiogenesis following SW treatment in vitro and in vivo [22, 32].

Associations of Mechanotransduction with Angiogenesis and Cell Proliferation
Mechanotransduction refers to the process in which vascular endothelial cells transmit extracellular mechanical stimuli (e.g., laminar shear stress) into intracellular signals that cause angiogenesis, cell proliferation, vasodilatation, and atheroprotection [33, 34]. Although mechanotransduction has been studied for decades, not until recently has the role of SW in mechanotransduction been investigated at the molecular level. It was previously observed that the expression of proliferating cell nuclear antigen is upregulated after SW treatment, suggesting that SW can promote cell proliferation [35]. It has been shown that mechanotransduction involves a variety of molecules, including VEGFR2, VE-cadherin (CD144), PECAM-1 (CD31), β1-integrin, mTOR, and cytoskeleton through which SW enhances angiogenesis and cell proliferation [21–23, 36] (Fig. 2).

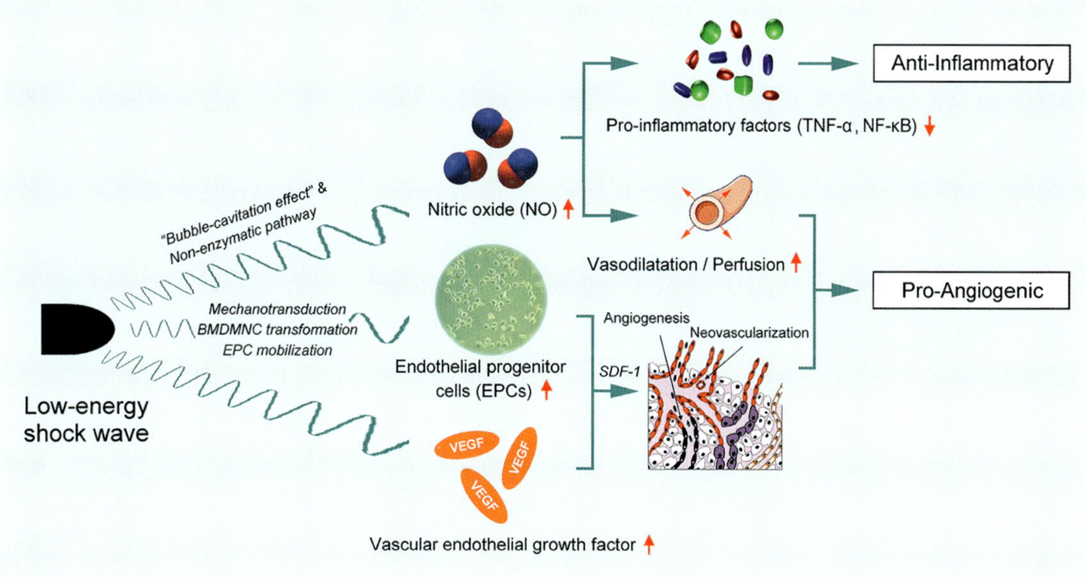

Fig. 1. Schematic illustration of anti-inflammatory and pro-angiogenic effects of extracorporeal shockwave on ischemic cardiovascular system. Nitric oxide (NO), which possesses both anti-inflammatory and vasodilatatory properties, is generated through shear stress-like "bubble-cavitation effect" through the action of endothelial nitric oxide synthase (eNOS) and also via the non-enzymatic pathway in the presence of hydrogen peroxide and physiological level of L-arginine. The number of circulating endothelial progenitor cells (EPCs) is elevated through shockwave-elicited mechanotransduction, transformation from bone marrow-derived mononuclear cells (BMDMNCs), and EPC mobilization. Shockwave also triggers the expression of stromal cell-derived factor 1 (SDF-1) that facilitates homing of EPCs to the ischemic site. Shockwave also induces the expression of vascular endothelial growth factors (VEGF) that contributes to angiogenesis.

Besides cell proliferation, cell homing has also been reported to contribute to SW-associated angiogenic effect. Experimental studies revealed that SW not only may improve tissue perfusion through transforming bone marrow-derived mononuclear cells (BMDMNCs) into EPCs [24], but it can also improve recruitment of circulating EPCs in SW-preconditioned ischemic tissue through the upregulation of the expression of chemoattractant factors [25] (Fig. 1). Consistently, one of our previous studies on a mini-pig chronic myocardial ischemia model showed that not only did SW therapy effectively reverse ischemia-elicited left ventricular dysfunction and remodeling through enhancing angiogenesis, suppressing inflammation, and oxidative stress, but it also enhanced the expressions of CXCR4 and stromal cell-derived factor-1α that suggest a role of SW in the regulation of stem/progenitor cell trafficking [31].

Combination of SW and Cell Therapy for Ischemic Cardiovascular Diseases
In concert with the ability of SW to enhance cell homing, it has been demonstrated that combined treatment with EPCs and SW is superior to either alone in improving

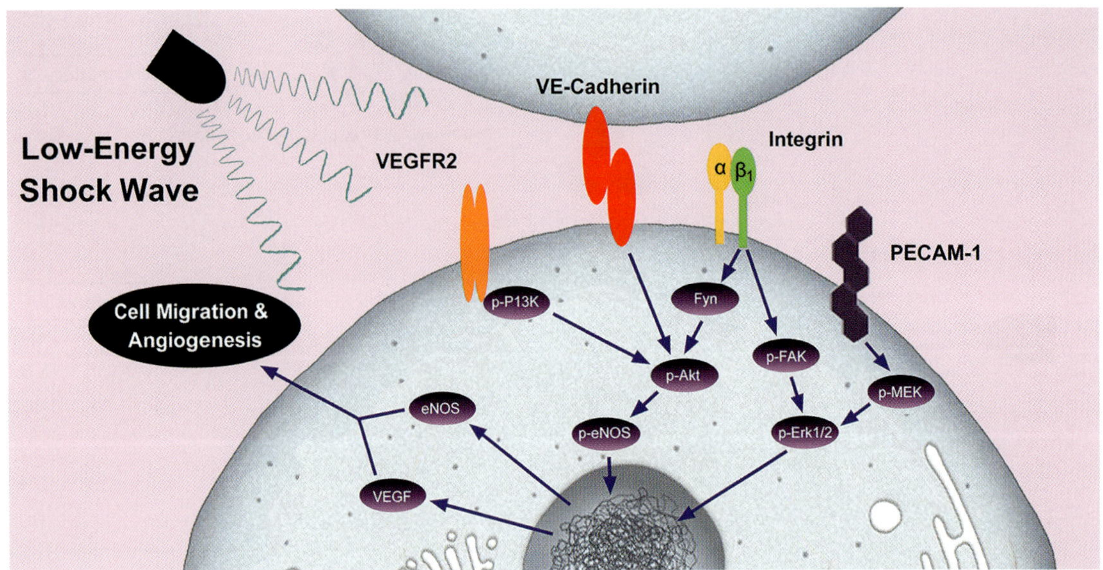

Fig. 2. Mechanisms underlying shockwave-induced angiogenesis through mechanotransduction. Low-energy shockwave (SW) causes the phosphorylation of Akt, eNOS, and Erk 1/2 in the process of mechanotransduction through the stimulation of the "mechanosensory complex" comprising VEG-FR-2, VE-cadherin, and PECAM-1. SW also enhances angiogenesis via triggering β1-integrin and its downstream pathways involving FAK, Erk1/2, Akt, and eNOS.

the survival of ischemic skin flaps in rats [37]. Besides, we have previously demonstrated that combined SW-BMDMSC therapy is superior to either alone for improving left ventricular function, reducing infarct size, and inhibiting left ventricular remodeling [38]. Interestingly, application of SW-pretreated BMDMNCs has also been found to be superior to BMDMNCs alone for preserving left ventricular function in a rabbit model of acute myocardial infarction. Furthermore, the recent CELLWAVE randomized clinical trial has shown that SW pretreatment targeted to the left ventricular anterior wall can reduce the incidence of major adverse cardiac events in patients with postinfarction chronic heart failure receiving intracoronary infusion of autologous BMDMNCs [39]. Therefore, it appears that combining SW and cell therapy, pretreatment of the cellular elements to be administered (i.e., EPCs, BMDMNCs) with SW, or application of SW to the target organ before cell therapy could all significantly improve the treatment outcome compared with that of monotherapy.

Conclusions

The mechanisms underlying extracorporeal SW treatment for ischemic cardiovascular diseases are multifold (Fig. 1, 2). Not only does it trigger the release of NO that possesses both vasodilatatory and anti-inflammatory actions, but it also upregulates the expression of VEGF that contributes to angiogenesis in the ischemic cardiovascular system. In addition to NO, SW also enables the release of a myriad of pro-angiogenic and anti-inflammatory molecules. Moreover, SW -elicited mechanotransduction enhances both angiogenesis and cell proliferation. SW may also improve perfusion in ischemic tissue by increasing circulating EPCs and, at the same time, securing EPCs to their ischemic targets through its cell homing effect. Combining cell treatment and low-energy SW appears to be a promising therapeutic strategy for treating ischemic cardiovascular diseases.

References

1 Mariotto S, de Prati AC, Cavalieri E, Amelio E, Marlinghaus E, Suzuki H: Extracorporeal shock wave therapy in inflammatory diseases: molecular mechanism that triggers anti-inflammatory action. Curr Med Chem 2009;16:2366–2372.

2 Clemedson CJ, Criborn CO: Mechanical response of different parts of a living body to a high explosive shock wave impact. Am J Physiol 1955;181:471–476.

3 Clemedson CJ, Jonsson A, Pettersson H: Propagation of an air-transmitted shock wave in muscular tissue. Nature 1956;177:380–381.

4 Chaussy C, Schmiedt E, Jocham D, Brendel W, Forssmann B, Walther V: First clinical experience with extracorporeally induced destruction of kidney stones by shock waves. J Urol 1982;127:417–420.

5 Nishida T, Shimokawa H, Oi K, Tatewaki H, Uwatoku T, Abe K, Matsumoto Y, Kajihara N, Eto M, Matsuda T, Yasui H, Takeshita A, Sunagawa K: Extracorporeal cardiac shock wave therapy markedly ameliorates ischemia-induced myocardial dysfunction in pigs in vivo. Circulation 2004;110:3055–3061.

6 Yan X, Zeng B, Chai Y, Luo C, Li X: Improvement of blood flow, expression of nitric oxide, and vascular endothelial growth factor by low-energy shockwave therapy in random-pattern skin flap model. Ann Plast Surg 2008;61:646–653.

7 Al-Abbad H, Simon JV: The effectiveness of extracorporeal shock wave therapy on chronic achilles tendinopathy: a systematic review. Foot Ankle Int 2013;34:33–41.

8 Wang CJ: Extracorporeal shockwave therapy in musculoskeletal disorders. J Orthop Surg Res 2012;7:11.

9 Fukumoto Y, Ito A, Uwatoku T, Matoba T, Kishi T, Tanaka H, Takeshita A, Sunagawa K, Shimokawa H: Extracorporeal cardiac shock wave therapy ameliorates myocardial ischemia in patients with severe coronary artery disease. Coron Artery Dis 2006;17:63–70.

10 Kikuchi Y, Ito K, Ito Y, Shiroto T, Tsuburaya R, Aizawa K, Hao K, Fukumoto Y, Takahashi J, Takeda M, Nakayama M, Yasuda S, Kuriyama S, Tsuji I, Shimokawa H: Double-blind and placebo-controlled study of the effectiveness and safety of extracorporeal cardiac shock wave therapy for severe angina pectoris. Circ J 2010;74:589–591.

11 Schmid JP, Capoferri M, Wahl A, Eshtehardi P, Hess OM: Cardiac shock wave therapy for chronic refractory angina pectoris. A prospective placebo-controlled randomized trial. Cardiovasc Ther 2013;31:e1–e6.

12 Ito K, Fukumoto Y, Shimokawa H: Extracorporeal shock wave therapy for ischemic cardiovascular disorders. Am J Cardiovasc Drugs 2011;11:295–302.

13 Ciampa AR, de Prati AC, Amelio E, Cavalieri E, Persichini T, Colasanti M, Musci G, Marlinghaus E, Suzuki H, Mariotto S: Nitric oxide mediates anti-inflammatory action of extracorporeal shock waves. FEBS Lett 2005;579:6839–6845.

14 Mariotto S, Cavalieri E, Amelio E, Ciampa AR, de Prati AC, Marlinghaus E, Russo S, Suzuki H: Extracorporeal shock waves: from lithotripsy to anti-inflammatory action by no production. Nitric Oxide 2005;12:89–96.

15 Oi K, Fukumoto Y, Ito K, Uwatoku T, Abe K, Hizume T, Shimokawa H: Extracorporeal shock wave therapy ameliorates hindlimb ischemia in rabbits. Tohoku J Exp Med 2008;214:151–158.

16 Zimpfer D, Aharinejad S, Holfeld J, Thomas A, Dumfarth J, Rosenhek R, Czerny M, Schaden W, Gmeiner M, Wolner E, Grimm M: Direct epicardial shock wave therapy improves ventricular function and induces angiogenesis in ischemic heart failure. J Thorac Cardiovasc Surg 2009;137:963–970.

17 Ito K, Fukumoto Y, Shimokawa H: Extracorporeal shock wave therapy as a new and non-invasive angiogenic strategy. Tohoku J Exp Med 2009;219:1–9.

18 Shao PL, Chiu CC, Yuen CM, Chua S, Chang LT, Sheu JJ, Sun CK, Wu CJ, Wang CJ, Yip HK: Shock wave therapy effectively attenuates inflammation in rat carotid artery following endothelial denudation by balloon catheter. Cardiology 2010;115:130–144.

19 Meirer R, Brunner A, Deibl M, Oehlbauer M, Piza-Katzer H, Kamelger FS: Shock wave therapy reduces necrotic flap zones and induces VEGF expression in animal epigastric skin flap model. J Reconstr Microsurg 2007;23:231–236.

20 Stojadinovic A, Elster EA, Anam K, Tadaki D, Amare M, Zins S, Davis TA: Angiogenic response to extracorporeal shock wave treatment in murine skin isografts. Angiogenesis 2008;11:369–380.

21 Hatanaka K, Ito K, Shindo T, Kagaya Y, Ogata T, Eguchi K, Kurosawa R, Shimokawa H: Molecular mechanisms of the angiogenic effects of low-energy shock wave therapy: roles of mechanotransduction. Am J Physiol Cell Physiol 2016;311:C378–C385.

22 Huang TH, Sun CK, Chen YL, Wang CJ, Yin TC, Lee MS, Yip HK: Shock wave enhances angiogenesis through VEGFR2 activation and recycling. Mol Med 2016;22, Epub ahead of print.

23 Lee FY, Zhen YY, Yuen CM, Fan R, Chen YT, Sheu JJ, Chen YL, Wang CJ, Sun CK, Yip HK: The mtor-fak mechanotransduction signaling axis for focal adhesion maturation and cell proliferation. Am J Transl Res 2017;9:1603–1617.

24 Yip HK, Chang LT, Sun CK, Youssef AA, Sheu JJ, Wang CJ: Shock wave therapy applied to rat bone marrow-derived mononuclear cells enhances formation of cells stained positive for cd31 and vascular endothelial growth factor. Circ J 2008;72:150–156.

25 Aicher A, Heeschen C, Sasaki K, Urbich C, Zeiher AM, Dimmeler S: Low-energy shock wave for enhancing recruitment of endothelial progenitor cells: a new modality to increase efficacy of cell therapy in chronic hind limb ischemia. Circulation 2006;114:2823–2830.

26 Li YS, Haga JH, Chien S: Molecular basis of the effects of shear stress on vascular endothelial cells. J Biomech 2005;38:1949–1971.

27 Maisonhaute E, Prado C, White PC, Compton RG: Surface acoustic cavitation understood via nanosecond electrochemistry. Part iii: shear stress in ultrasonic cleaning. Ultrason Sonochem 2002;9:297–303.

28 Gotte G, Amelio E, Russo S, Marlinghaus E, Musci G, Suzuki H: Short-time non-enzymatic nitric oxide synthesis from l-arginine and hydrogen peroxide induced by shock waves treatment. FEBS Lett 2002;520:153–155.

29 Kuo YR, Wu WS, Hsieh YL, Wang FS, Wang CT, Chiang YC, Wang CJ: Extracorporeal shock wave enhanced extended skin flap tissue survival via increase of topical blood perfusion and associated with suppression of tissue pro-inflammation. J Surg Res 2007;143:385–392.

30 Davis TA, Stojadinovic A, Anam K, Amare M, Naik S, Peoples GE, Tadaki D, Elster EA: Extracorporeal shock wave therapy suppresses the early proinflammatory immune response to a severe cutaneous burn injury. Int Wound J 2009;6:11–21.

31 Fu M, Sun CK, Lin YC, Wang CJ, Wu CJ, Ko SF, Chua S, Sheu JJ, Chiang CH, Shao PL, Leu S, Yip HK: Extracorporeal shock wave therapy reverses ischemia-related left ventricular dysfunction and remodeling: molecular-cellular and functional assessment. PLoS One 2011;6:e24342.

32 Holfeld J, Tepekoylu C, Blunder S, Lobenwein D, Kirchmair E, Dietl M, Kozaryn R, Lener D, Theurl M, Paulus P, Kirchmair R, Grimm M: Low energy shock wave therapy induces angiogenesis in acute hind-limb ischemia via VEGF receptor 2 phosphorylation. PLoS One 2014;9:e103982.

33 Hahn C, Schwartz MA: Mechanotransduction in vascular physiology and atherogenesis. Nat Rev Mol Cell Biol 2009;10:53–62.

34 Traub O, Berk BC: Laminar shear stress: mechanisms by which endothelial cells transduce an atheroprotective force. Arterioscler Thromb Vasc Biol 1998;18:677–685.

35 Kuo YR, Wang CT, Wang FS, Chiang YC, Wang CJ: Extracorporeal shock-wave therapy enhanced wound healing via increasing topical blood perfusion and tissue regeneration in a rat model of STZ-induced diabetes. Wound Repair Regen 2009;17:522–530.

36 Ha CH, Kim S, Chung J, An SH, Kwon K: Extracorporeal shock wave stimulates expression of the angiogenic genes via mechanosensory complex in endothelial cells: mimetic effect of fluid shear stress in endothelial cells. Int J Cardiol 2013;168:4168–4177.

37 Zhang X, Yan X, Wang C, Lu S, Tang T, Chai Y: The effect of autologous endothelial progenitor cell transplantation combined with extracorporeal shock-wave therapy on ischemic skin flaps in rats. Cytotherapy 2014;16:1098–1109.

38 Sheu JJ, Lee FY, Yuen CM, Chen YL, Huang TH, Chua S, Chen CH, Chai HT, Sung PH, Chang HW, Sun CK, Yip HK: Combined therapy with shock wave and autologous bone marrow-derived mesenchymal stem cells alleviates left ventricular dysfunction and remodeling through inhibiting inflammatory stimuli, oxidative stress & enhancing angiogenesis in a swine myocardial infarction model. Int J Cardiol 2015;193:69–83.

39 Assmus B, Walter DH, Seeger FH, Leistner DM, Steiner J, Ziegler I, Lutz A, Khaled W, Klotsche J, Tonn T, Dimmeler S, Zeiher AM: Effect of shock wave-facilitated intracoronary cell therapy on lvef in patients with chronic heart failure: the cellwave randomized clinical trial. JAMA 2013;309:1622–1631.

Hon-Kan Yip, MD
Division of Thoracic and Cardiology, Kaohsiung Chang Gung Memorial Hospital
123 Tai-Pei Road, Niao Sung District
Kaohsiung 83301 (Taiwan)
E-Mail han.gung@msa.hinet.net

Wang C-J, Schaden W, Ko J-Y (eds): Shockwave Medicine.
Transl Res Biomed. Basel, Karger, 2018, vol 6, pp 109–116 (DOI: 10.1159/000485068)

Effect of Extracorporeal Shockwave on Angiogenesis and Anti-Inflammation: Molecular-Cellular Signaling Pathways

Steve Leu[a] · Tien-Hung Huang[b] · Yi-Ling Chen[b] · Hon-Kan Yip[a–c]

[a]Institute for Translational Research in Biomedicine, [b]Division of Cardiology, Department of Internal Medicine, and [c]Center for Shockwave Medicine and Tissue Engineering, Kaohsiung Chang Gung Memorial Hospital and Chang Gung University College of Medicine, Kaohsiung, Taiwan

Abstract

Low energy extracorporeal shockwave (SW) therapy (ESWT) has been used to improve blood flow recovery and reduce inflammation in muscle and organs with ischemia/reperfusion injury. In recent decades, through cultured cell models and experimental animal studies, several intra-cellular signaling transductions and inter-cellular communications, which participate in SW-induced angiogenesis and anti-inflammation, were identified. Purinergic receptors, cytoskeletal proteins, and intracellular trafficking systems are found to be responsible for mechanotransduction of SW. Downstream signaling of SW, including Akt and extracellular signal-regulated kinase (ERK1/2), is indicated to regulate expression and activation of angiogenesis-associated protein endothelial nitric oxide synthase and VEGF. Additionally, the degradation and recycling of vascular endothelial growth factor receptor in endothelial cells are also regulated by SW. Furthermore, ESWT also regulates the concentration of chemokines and then enhances the recruitment of endothelial progenitor cells to the injury region for revascularization. In addition to angiogenesis, the anti-inflammation effect of ESWT also plays important role in recovery from injury. Through immunomodulation, ESWT regulates the inflammatory reactions, reduces infiltration of inflammatory cells, and promotes differentiation of M2 macrophages. Taken together, not only the cellular response triggered directly by mechanotransduction, the crosstalk among local injury cells, circulating endothelial progenitor cells, and infiltrated immune cells all participate in enhancing angiogenesis and anti-inflammation in ESWT.

Mechanotransduction of Shockwave

In contrast to high energy shockwave (SW) in lithotripsy, several biological effects, particularly in angiogenesis and inflammatory regulation-associated cellular responses are induced by low energy SW in various tissues and organs [1]. Although the therapeutic efficacy of extracorporeal SW therapy (ESWT) in ameliorating ischemia/reperfusion and regional inflammatory injury is convincing, the underlying mechanism, particularly the part regarding mechanotransduction, is still not completely clear.

In a study regarding stem cell differentiation, researchers showed that SW treatment triggered the release of adenosine 5'-triphosphate (ATP), followed by activation of the purinergic P2X7 receptor and intracellular p38-MAPK signaling, and then promoted the osteogenic differentiation of human mesenchymal stem cells [2]. In the process of SW-triggered osteogenic differentiation, the release of ATP from cytoplasm to extracellular space is considered as the first and most critical step to transduce physical SW to biological signal transduction [2]. The SW-triggered release of intracellular ATP was also observed in other types of cells. Weihs et al. [3] reported that SW could trigger ATP release, activate phosphorylation of extracellular signal-regulated kinase (ERK1/2), and enhance proliferation in murine mesenchymal stem cells, human adipose-derived mesenchymal stem cells (ADMSC), and immortal human Jurkat T cell line [3]. In an earlier study, Yu et al. [4] also demonstrated the release of ATP was associated with focal adhesion kinase (FAK) phosphorylation, IL-2 production, and cellular proliferation in T cells.

In addition to purinergic receptor-mediated signal transduction, cytoskeleton and intracellular trafficking system also participate in the mechanotransduction of SW [5, 6]. In an ADMSC model, Lee et al. [6] demonstrated that SW treatment induced FAK translocation to focal adhesion, enhanced phosphorylation of paxillin, rendered α-actinin redistribution of focal adhesion, and promoted cell proliferation. Moreover, the activation of FAK signaling in SW-treated ADMSC depended on the activation of mechanistic target of rapamycin (mTOR) signaling and integrity of microfilaments. Of importance, excessive SW resulted in cytoskeletal damage and then inhibited the activation of mTOR and FAK signaling [6]. In a human umbilical vein endothelial cell (HUVEC) model, researchers observed that SW not only induced the phosphorylation of Akt and ERK1/2, but also increased the phosphorylation level of caveolin-1 and expression of HUTS-4 which represents β1-integrin activity [5]. Furthermore, knockdown of caveolin-1 and β1-integrin abolished the activation of Akt and ERK1/2, as well as the SW-induced enhancement of angiogenic activity in HUVEC [5]. Caveolin-1 and β1-integrin have found to be involved in the endocytosis and intracellular trafficking, indicating that the membrane protein and intracellular trafficking may also have a role in mechanotransduction of SW.

Angiogenic Signaling Regulated by ESWT in Endothelial Cells

After ischemic injury occurs, to reestablish the blood flow is a critical step to ameliorate tissue damage and preserve organ function. Hence, the effect of ESWT on enhancing angiogenesis is particularly worthy of note. In preclinical animal models with

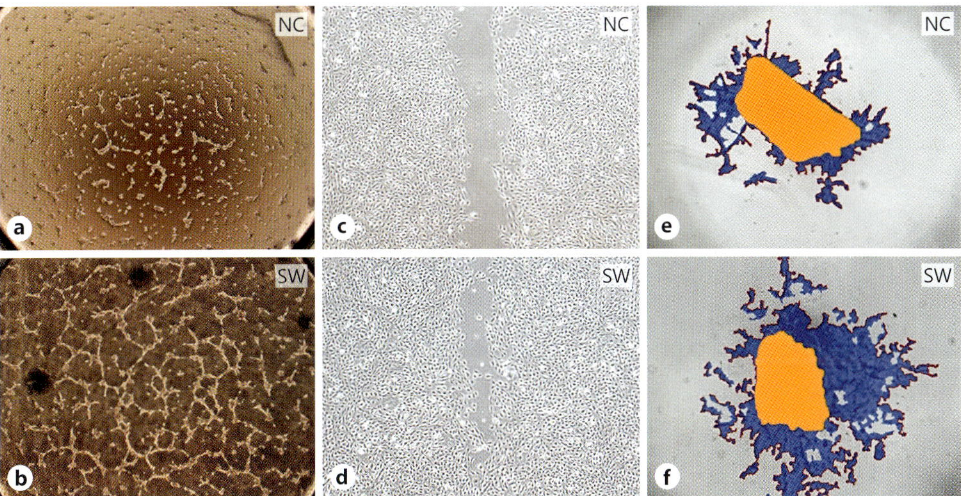

Fig. 1. In vitro and ex vivo studies to validate the effect of shockwave in enhancing angiogenesis. In vitro and ex vivo angiogenic assays were applied to examine the effect of shockwave (SW, 0.12 mJ/mm^2, 200 pulses, 2 Hz) in modulating angiogenic activity. **a, b** Matrigel-based tube formation assay in HUVECs. **c, d** Migration assay with scratch wound healing in HUVECs. **e, f** Endothelial sprouting assay in rat carotid artery. **a, c, e** Cells or tissues without SW treatment. **b, d, f** Cells or tissues with SW treatment. The effect of SW on enhancing angiogenesis was identified through the enhancement of tube formation (**b**), would healing (**d**), and endothelial sprouting (**f**). NC, normal control.

critical limb ischemia and ischemic heart injury, ESWT was shown to have the ability to improve blood flow recovery through enhancing angiogenesis [7, 8]. Not only in animal studies, the clinical practice of ESWT has also been applied to patients with ischemic heart disease [9, 10]. The interaction between endothelial cells and their environment is involved in the process of angiogenesis, while angiogenic signalings, such as vascular endothelial growth factor (VEGF), regulates the motility, sprouting, and proliferation of endothelial cells [11, 12]. Results from in vitro and ex vivo experiments have demonstrated that SW modulated the angiogenic signaling and activity in endothelial cells (Fig. 1). Several studies have demonstrated that the expression levels of VEGF in endothelial cells can be increased by ESWT through transcriptional regulation [5, 13, 14]. Following the binding of VEGF to its receptors, the downstream ERK signaling and Akt- endothelial nitric oxide synthase (eNOS) activation mediated the SW-triggered migration, sprouting, and tube formation of endothelial cells [5, 15]. Hatanaka et al. [5] indicated that the mRNA and protein expression level of eNOS, which participates in anti-inflammation and angiogenesis, was increased by low energy SW (800 pulses, 0.03 mJ/mm^2, 1 Hz) in HUVECs. In addition to the synthesis and release of VEGF, the expression levels of VEGF receptors are also regulated by SW [13, 15, 16]. The protein and mRNA expression level of VEGF receptor 1 (VEGFR1, Flt-1) and 2 (VEFGR2, KDR) were found to be upregulated in porcine myocardial infarction and murine model with critical limb ischemia [13, 16]. Recently, Huang et al. [15] also indicated the VEGFR2-Akt-eNOS axis regulated the generation of endogenous nitric oxide (NO), cellular migration, endothelial sprouting, as well as enhanced blood flow recovery in a rat model with critical limb ischemia. The recycling of VEGFR2 was associated with intracellular trafficking and important in modulating activity of this VEGFR2-Akt-eNOS axis.

So far, most studies with investigation on mechanotransduction of SW indicated that FAK played a pivotal role to transduce SW in cells. The participation of FAK in physiological or pathological angiogenesis was reported in various studies, particularly in the VEGF-induced angiogenesis in endothelial cells [17–19]. In physiological condition, VEGF is synthesized and released from non-endothelial cells. SW was found to increase the expression level of VEGF in several types of non-endothelial cells, including fibroblast, cardiomyocytes, and neural cells (neuron, astrocyte, and oligodendrocytes) and then promote angiogenesis in the nearby area [20–22]. In HU-VECs, the knockdown of FAK reduced the mRNA expression level of SW-upregulated VEGF [5]. Hence, it is reasonable that, during the process of ESWT, SW enhances angiogenesis and improve blood flow recovery through upregulating the synthesis and release of VEGF (in endothelial and non-endothelial cells), enhancing the expression and recycling of VEGFR in endothelial cells, and activating intracellular FAK, Akt, ERK signalings for transcriptional and post-translational regulation of angiogenic factors.

ESWT Enhances the Recruitment of Endothelial Progenitor Cells

In normal conditions, endothelial progenitor cells (EPCs) resides in the bone marrow. During the ischemic condition, EPCs migrate from bone marrow to circulation and eventually to the ischemic region and then differentiate into endothelial cells for neovascularization [23]. Recent studies demonstrated the number and quality of circulating EPCs is highly relevant to the prognosis of ischemic organ injury, that is, myocardial infarction [24]. Hence, not only as an important issue in the field of stem cell research, to enhance the survival, endothelial differentiation, angiogenic activity and recruitment of EPCs should be an effective strategy to enhance the therapeutic efficacy of stem cell therapy against ischemic injury. Several studies have showed that SW treatment could reduce oxidative stress, a common physiological insult, which decreases the number and angiogenic capacity of EPCs in organs with ischemia/reperfusion injury [21, 25, 26]. Furthermore, in the murine model, low energy SW treatment enhanced the recruitment of EPCs in the ischemic muscles [27]. Hence, a combination of stem cells and ESWT should be an effective strategy to treat ischemia/reperfusion injury.

Sheu et al. [26] applied a porcine model to examine the therapeutic efficacy of combined therapy with ESWT and bone marrow-derived mesenchymal stem cell (BMDMSC) against myocardial infarction. Results showed both ESWT and BMD-

MSC reduced the myocardial infarction and preserved cardiac function through inhibiting inflammation, ameliorating oxidative stress, and enhancing angiogenesis [26]. It is worthy to note that the most functional enhancement and structure preservation were observed in those treated with combined therapy. The synergistic effect in combination of ESWT and stem cell therapy in ameliorating ischemic injury and improving blood flow recovery was also observed in a rat model with critical limb ischemia [28]. The retention of transplanted bone marrow-derived endothelial progenitor cells (BMDEPCs) in ischemic quadriceps was increased in CLI rats that received ESWT in the ischemic limb. The increased number of transplanted BMDEPCs by ESWT was observed in the ischemic quadriceps was along with the increased expression level of stromal cell-derived factor-1α, a chemokine that attracts CXCR4+ BMDECPS migrating from circulation to ischemic tissues [28]. Both studies demonstrated that ESWT not only directly activated the angiogenic signaling in ischemic area, it also enhanced neovascularization and improved the blood flow recovery through increasing generation of chemokines to recruit stem cells.

Immunomodulation by ESWT

Excessive number of high energy SW lithotripsy was found to cause acute oxidative stress and severe inflammation in renal parenchyma [29]. In contrast to enhancing angiogenesis by low energy SW, excessive SW renders HUVEC cells to an endothelial pro-inflammatory phenotype [30]. However, the low energy SW with appropriate power and pulses is indicated with benefit in reducing inflammatory reactions [25, 31–33]. As that involved in the ESWT-induced angiogenesis, local cellular response, recruitment of circulating immune cell, and interactions between immune cells and damaged cells all participate in modulating inflammatory reactions by ESWT. In vitro study with C6 glioma cells showed that low energy SW could increase the expression level of nNOS, which is responsible for synthesizing "physiological" amount of NO to maintain fundamental events such as retrograde neurotransmission, long-terminal potentiation, angiogenesis, and immunomodulation [33, 34]. In endothelial cells, SW also increases the expression level of eNOS, which has similar function to generate physiological amount of NO to maintain endothelial function. Of importance, the appropriate physiological level of NO strongly suppresses the activation of nuclear factor (NF)-κB [33, 34]. Hence, it is reasonable that the effect of SW in anti-inflammation is through increasing expression level of eNOS/nNOS to maintain homeostasis of NO for suppressing pro-inflammatory NF-κB signaling and release of inflammatory cytokines.

Additionally, the recruitment of immune cells also participates in the inflammatory response after ischemia/reperfusion injury. In a large animal model study, porcine with acute myocardial infarction was applied to investigate the efficacy of ESWT in ameliorating myocardial injury. Result showed that the number of infiltrated neutrophils and macrophages in myocardium was reduced by ESWT [31]. Not only in myocardium, a reduction of infiltrated neutrophils and macrophages was also observed in severe cutaneous burn injury, arterial endothelial denudation, and acute interstitial cystitis [25, 32, 35].

In addition to the number of infiltrated immune cells, the population or polarization of infiltrated immune cells also plays a pivotal role in modulating inflammatory responses after SW treatment. In general, macrophages can polarize into M1 and M2 macrophages that have distinct functional in regulation inflammatory reactions. Evidences from a mouse model with critical limb ischemia showed that ESWT increase the expression of the M2 polarization promoting chemokine interleukin (IL)-13 and mRNA expression of the M2 scavenger receptor CD163, indicating the polarization of anti-inflammatory M2 macrophage was enhanced by ESWT [36]. On the other hand, directly applying low energy SW on human monocyte-derived macrophages showed similar results. The expression levels of pro-inflammatory M1 macrophage markers (CD80, COX2, CCL5) was reduced by low energy SW, while the expression level of M2 macrophage marker was increased [37]. Along with the changes in M1/M2 marker expression levels, the production of inflammatory cytokine IL-1β was decreased, while the expression level of anti-inflammatory cytokine IL-10 was elevated [37]. Taken together, by inhibiting NF-κB signaling, decreasing cytokine synthesis and release, reducing macrophages and neutrophils infiltration, and promoting M2 macrophage polarization, the inflammatory stimuli in organs with ischemia/reperfusion injury could be reduced by ESWT.

In conclusion, cumulating evidences suggest that ESWT has the therapeutic efficacy in ameliorating inflammatory and ischemia/reperfusion injury by enhancing angiogenesis and immunomodulation. In physiological conditions, ESWT-triggered angiogenesis and immune regulation are mediated by the participation of onsite cells/tissues and recruited EPCs and immune cells. Several membranes or cytoskeletal-associated proteins, including P2X7, FAK, caveolin-1, and intergrin-1β, are considered mechanotransduction sensors to directly transmit physical forces to biological signalings. However, more studies are needed to fully reveal the underlying mechanism of SW in modulating angiogenesis and inflammation.

Disclosure Statement

The authors declared that they did not receive any honoraria or consultancy fees in the writing this manuscript. No benefits in any form have been received or will be received from a commercial party related directly or indirectly to the subject of this article.

Leu · Huang · Chen · Yip

References

1 Sun CK, Shao PL, Wang CJ, Yip HK: Study of vascular injuries using endothelial denudation model and the therapeutic application of shock wave: a review. Am J Transl Res 2011;3:259–268.

2 Sun D, Junger WG, Yuan C, Zhang W, Bao Y, Qin D, Wang C, Tan L, Qi B, Zhu D, Zhang X, Yu T: Shockwaves induce osteogenic differentiation of human mesenchymal stem cells through atp release and activation of P2X7 receptors. Stem Cells 2013;31:1170–1180.

3 Weihs AM, Fuchs C, Teuschl AH, Hartinger J, Slezak P, Mittermayr R, Redl H, Junger WG, Sitte HH, Runzler D: Shock wave treatment enhances cell proliferation and improves wound healing by atp release-coupled extracellular signal-regulated kinase (ERK) activation. J Biol Chem 2014;289:27090–27104.

4 Yu T, Junger WG, Yuan C, Jin A, Zhao Y, Zheng X, Zeng Y, Liu J: Shockwaves increase T-cell proliferation and IL-2 expression through ATP release, P2X7 receptors, and FAK activation. Am J Physiol Cell Physiol 2010;298:C457–C464.

5 Hatanaka K, Ito K, Shindo T, Kagaya Y, Ogata T, Eguchi K, Kurosawa R, Shimokawa H: Molecular mechanisms of the angiogenic effects of low-energy shock wave therapy: roles of mechanotransduction. Am J Physiol Cell Physiol 2016;311:C378–C385.

6 Lee FY, Zhen YY, Yuen CM, Fan R, Chen YT, Sheu JJ, Chen YL, Wang CJ, Sun CK, Yip HK: The mtor-fak mechanotransduction signaling axis for focal adhesion maturation and cell proliferation. Am J Transl Res 2017;9:1603–1617.

7 Holfeld J, Zimpfer D, Albrecht-Schgoer K, Stojadinovic A, Paulus P, Dumfarth J, Thomas A, Lobenwein D, Tepekoylu C, Rosenhek R, Schaden W, Kirchmair R, Aharinejad S, Grimm M: Epicardial shock-wave therapy improves ventricular function in a porcine model of ischaemic heart disease. J Tissue Eng Regen Med 2016;10:1057–1064.

8 Lei PP, Tao SM, Shuai Q, Bao YX, Wang SW, Qu YQ, Wang DH: Extracorporeal cardiac shock wave therapy ameliorates myocardial fibrosis by decreasing the amount of fibrocytes after acute myocardial infarction in pigs. Coron Artery Dis 2013;24:509–515.

9 Ito K, Fukumoto Y, Shimokawa H: Extracorporeal shock wave therapy as a new and non-invasive angiogenic strategy. Tohoku J Exp Med 2009;219:1–9.

10 Zhao L, Yang P, Tang Y, Li R, Peng Y, Wang Y, Pu L, Guo T: Effect of cardiac shock wave therapy on the microvolt t wave alternans of patients with coronary artery disease. Int J Clin Exp Med 2015;8:16463–16471.

11 Jakobsson L, Franco CA, Bentley K, Collins RT, Ponsioen B, Aspalter IM, Rosewell I, Busse M, Thurston G, Medvinsky A, Schulte-Merker S, Gerhardt H: Endothelial cells dynamically compete for the tip cell position during angiogenic sprouting. Nat Cell Biol 2010;12:943–953.

12 Eilken HM, Adams RH: Dynamics of endothelial cell behavior in sprouting angiogenesis. Curr Opin Cell Biol 2010;22:617–625.

13 Nishida T, Shimokawa H, Oi K, Tatewaki H, Uwatoku T, Abe K, Matsumoto Y, Kajihara N, Eto M, Matsuda T, Yasui H, Takeshita A, Sunagawa K: Extracorporeal cardiac shock wave therapy markedly ameliorates ischemia-induced myocardial dysfunction in pigs in vivo. Circulation 2004;110:3055–3061.

14 Peng YZ, Zheng K, Yang P, Wang Y, Li RJ, Li L, Pan JH, Guo T: Shock wave treatment enhances endothelial proliferation via autocrine vascular endothelial growth factor. Genet Mol Res 2015;14:19203–19210.

15 Huang TH, Sun CK, Chen YL, Wang CJ, Yin TC, Lee MS, Yip HK: Shock wave enhances angiogenesis through VEGFR2 activation and recycling. Mol Med 2016;22, Epub ahead of print.

16 Holfeld J, Tepekoylu C, Blunder S, Lobenwein D, Kirchmair E, Dietl M, Kozaryn R, Lener D, Theurl M, Paulus P, Kirchmair R, Grimm M: Low energy shock wave therapy induces angiogenesis in acute hind-limb ischemia via vegf receptor 2 phosphorylation. PLoS One 2014;9:e103982.

17 Cabrita MA, Jones LM, Quizi JL, Sabourin LA, McKay BC, Addison CL: Focal adhesion kinase inhibitors are potent anti-angiogenic agents. Mol Oncol 2011;5:517–526.

18 Tavora B, Batista S, Reynolds LE, Jadeja S, Robinson S, Kostourou V, Hart I, Fruttiger M, Parsons M, Hodivala-Dilke KM: Endothelial fak is required for tumour angiogenesis. EMBO Mol Med 2010;2:516–528.

19 Zhao X, Guan JL: Focal adhesion kinase and its signaling pathways in cell migration and angiogenesis. Adv Drug Deliv Rev 2011;63:610–615.

20 Fu M, Sun CK, Lin YC, Wang CJ, Wu CJ, Ko SF, Chua S, Sheu JJ, Chiang CH, Shao PL, Leu S, Yip HK: Extracorporeal shock wave therapy reverses ischemia-related left ventricular dysfunction and remodeling: molecular-cellular and functional assessment. PLoS One 2011;6:e24342.

21 Yuen CM, Chung SY, Tsai TH, Sung PH, Huang TH, Chen YL, Chen YL, Chai HT, Zhen YY, Chang MW, Wang CJ, Chang HW, Sun CK, Yip HK: Extracorporeal shock wave effectively attenuates brain infarct volume and improves neurological function in rat after acute ischemic stroke. Am J Transl Res 2015;7:976–994.

22 Frairia R, Berta L: Biological effects of extracorporeal shock waves on fibroblasts. A review. Muscles Ligaments Tendons J 2011;1:138–147.

23 Asahara T, Masuda H, Takahashi T, Kalka C, Pastore C, Silver M, Kearne M, Magner M, Isner JM: Bone marrow origin of endothelial progenitor cells responsible for postnatal vasculogenesis in physiological and pathological neovascularization. Circ Res 1999;85:221–228.

24 Hill JM, Zalos G, Halcox JP, Schenke WH, Waclawiw MA, Quyyumi AA, Finkel T: Circulating endothelial progenitor cells, vascular function, and cardiovascular risk. N Engl J Med 2003;348:593–600.

25 Chen YT, Yang CC, Sun CK, Chiang HJ, Chen YL, Sung PH, Zhen YY, Huang TH, Chang CL, Chen HH, Chang HW, Yip HK: Extracorporeal shock wave therapy ameliorates cyclophosphamide-induced rat acute interstitial cystitis though inhibiting inflammation and oxidative stress – in vitro and in vivo experiment studies. Am J Transl Res 2014;6:631–648.

26 Sheu JJ, Lee FY, Yuen CM, Chen YL, Huang TH, Chua S, Chen YL, Chen CH, Chai HT, Sung PH, Chang HW, Sun CK, Yip HK: Combined therapy with shock wave and autologous bone marrow-derived mesenchymal stem cells alleviates left ventricular dysfunction and remodeling through inhibiting inflammatory stimuli, oxidative stress & enhancing angiogenesis in a swine myocardial infarction model. Int J Cardiol 2015;193:69–83.

27 Aicher A, Heeschen C, Sasaki K, Urbich C, Zeiher AM, Dimmeler S: Low-energy shock wave for enhancing recruitment of endothelial progenitor cells: a new modality to increase efficacy of cell therapy in chronic hind limb ischemia. Circulation 2006;114:2823–2830.

28 Yeh KH, Sheu JJ, Lin YC, Sun CK, Chang LT, Kao YH, Yen CH, Shao PL, Tsai TH, Chen YL, Chua S, Leu S, Yip HK: Benefit of combined extracorporeal shock wave and bone marrow-derived endothelial progenitor cells in protection against critical limb ischemia in rats. Crit Care Med 2012;40:169–177.

29 Clark DL, Connors BA, Evan AP, Handa RK, Gao S: Effect of shock wave number on renal oxidative stress and inflammation. BJU Int 2011;107:318–322.

30 Sonden A, Johansson AS, Palmblad J, Kjellstrom BT: Proinflammatory reaction and cytoskeletal alterations in endothelial cells after shock wave exposure. J Invest Med 2006;54:262–271.

31 Abe Y, Ito K, Hao K, Shindo T, Ogata T, Kagaya Y, Kurosawa R, Nishimiya K, Satoh K, Miyata S, Kawakami K, Shimokawa H: Extracorporeal low-energy shock-wave therapy exerts anti-inflammatory effects in a rat model of acute myocardial infarction. Circ J 2014;78:2915–2925.

32 Davis TA, Stojadinovic A, Anam K, Amare M, Naik S, Peoples GE, Tadaki D, Elster EA: Extracorporeal shock wave therapy suppresses the early proinflammatory immune response to a severe cutaneous burn injury. Int Wound J 2009;6:11–21.

33 Mariotto S, de Prati AC, Cavalieri E, Amelio E, Marlinghaus E, Suzuki H: Extracorporeal shock wave therapy in inflammatory diseases: molecular mechanism that triggers anti-inflammatory action. Curr Med Chem 2009;16:2366–2372.

34 Ciampa AR, de Prati AC, Amelio E, Cavalieri E, Persichini T, Colasanti M, Musci G, Marlinghaus E, Suzuki H, Mariotto S: Nitric oxide mediates anti-inflammatory action of extracorporeal shock waves. FEBS Lett 2005;579:6839–6845.

35 Shao PL, Chiu CC, Yuen CM, Chua S, Chang LT, Sheu JJ, Sun CK, Wu CJ, Wang CJ, Yip HK: Shock wave therapy effectively attenuates inflammation in rat carotid artery following endothelial denudation by balloon catheter. Cardiology 2010;115:130–144.

36 Tepekoylu C, Lobenwein D, Urbschat A, Graber M, Pechriggl EJ, Fritsch H, Paulus P, Grimm M, Holfeld J: Shock wave treatment after hindlimb ischaemia results in increased perfusion and m2 macrophage presence. J Tissue Eng Regen Med 2016, Epub ahead of print.

37 Sukubo NG, Tibalt E, Respizzi S, Locati M, d'Agostino MC: Effect of shock waves on macrophages: a possible role in tissue regeneration and remodeling. Int J Surg 2015;24:124–130.

Hon-Kan Yip, MD
Division of Cardiology, Department of Internal Medicine, Chang Gung Memorial Hospital
123, Ta Pei Road, Niaosong Dist.
Kaohsiung 83301 (Taiwan)
E-Mail han.gung@msa.hinet.net

Wang C-J, Schaden W, Ko J-Y (eds): Shockwave Medicine.
Transl Res Biomed. Basel, Karger, 2018, vol 6, pp 117–126 (DOI: 10.1159/000485069)

Extracorporeal Shockwave Therapy Assisted Intravesical Drug Delivery

Pradeep Tyagi[a] · Yao-Chi Chuang[b,c]

[a]Department of Urology, University of Pittsburgh School of Medicine, Pittsburgh, PA, USA; [b]Department of Urology and [c]Center for Shockwave Medicine and Tissue Engineering, Kaohsiung Chang Gung Memorial Hospital, Chang Gung University College of Medicine, Kaohsiung, Taiwan

Abstract

High-energy extracorporeal shockwaves have been used for over 30 years to crush upper urinary tract stones. Low energy shockwave (LESW) has been clinically used to improve tissue regeneration at tendon-bone junction, ischemic cardiovascular disorders, and erectile dysfunction. Furthermore, shockwave has been shown to temporarily increase tissue permeability through micro- and submicrosized deformations lasting for nanosecond to a micro second time scale, which can facilitate to deliver macromolecules or drug into cells without cytotoxicity. Recently, there has been a lot of interest in exploring whether abrupt perturbations in urothelium from shockwaves traversing through bladder can temporarily reduce the impermeability barrier erected for large molecular weight drugs instilled into bladder. LESW can impale the cell membrane by a dint of its acoustic microcavitation effect. Previous preclinical study has shown that LESW potentiates the efficacy of intravesical botulinum toxin A pretreatment in suppressing the acetic acid induced bladder hyperactivity and inflammatory reaction. Therefore, LESW may be a promising alternative for intravesical BoNT-A or other drugs delivery.

© 2018 S. Karger AG, Basel

Introduction

Rationale for Intravesical Delivery
Although a large number of oral therapeutic agents are available for treating bladder diseases, only a small fraction of administered drug reaches the bladder lumen as a urine constituent owing to metabolism. One approach to circumvent this problem is to administer larger doses of therapeutic agents by systemic route, which is not the option for drugs with narrow therapeutic index [1]. Risk of side effects also increases with undesired rise in biodistribution.

Another approach is the intravesical drug delivery, which involves direct administration of toxic or chemotherapeutic drugs into the urinary bladder through a catheter. Delivering a drug close to the targeted cells improves its benefit versus risk ratio. Intravesical drug delivery is a first-line treatment for delaying or preventing the recurrence of bladder cancer [2, 3]. The anatomy of the urinary bladder allows relatively uncomplicated access and manipulation with a catheter, and hence localized treatment options have enormous potential. Given the success of this approach in oncology, the lessons learned over the decades in treating bladder cancer have been extended to improve the clinical management of noncancerous lower urinary tract disorder such as interstitial cystitis/painful bladder syndrome (IC/PBS) [4], radiation cystitis [5] and refractory overactive bladder [6].

Pros and Cons
The lining of urinary bladder also known as urothelium is highly impermeable [7, 8], which offers unique opportunities as well as challenges in drug delivery. Instillation of drugs in the bladder provides a high concentration of drugs locally at the disease site in the bladder without an increase in systemic levels, which can explain the low risk of systemic side effects. Delivery of a therapeutic agent directly into the bladder greatly improves the exposure of the affected bladder lining to the agent, and this is of even greater significance in the case of drug-resistant targets, which require a much-higher dose of the drug [9, 10]. Intravesical therapy is frequently used as adjunct to oral treatment regimens or as second-line treatment [11] in clinical management.

Impermeability of urothelium for large molecular weight drugs typically limits the opportunity for systemic distribution after instillation and leads to lower side effects [12]. Since most of the new drugs in urology are macromolecules, such as genes, proteins, which cannot penetrate into the urothelium. Therefore, the drug delivery to bladder tissues by the intravesical route is constrained by the impermeability of urothelial cells, a short duration of action, and the need for frequent administration [13–15].

Bladder Urothelium

The urothelium, responsible for a multitude of physiological activities, is not only effective in blocking the entry of urine contents but also equally effective in blocking the entry of instilled drugs [14]. It sits at the interface between the urine and underlying connective tissue where it forms a barrier preventing the unregulated exchange of solutes, ions, and toxic metabolites. The permeability to substances such as water, ammonia, and urea, which normally cross membranes relatively rapidly, is extremely low in the urothelium [14]. Urothelial tight junctions are comprised of a dense network of cytoplasmic proteins (zonula occludens-1), cytoskeletal elements, and transmembrane proteins (occludin and claudins) and have the highest recorded paracellular resistance of all epithelia measured to date [16]. In addition, the apical plasma membrane of the urothelium contains specific proteins, uroplakins, which account for the transcellular resistance. Electron microscopy of umbrella cells lying at the luminal surface of the urothelium showed a hexagonal arrangement of uroplakins, where 6 subunits of each particle are tightly joined to form a complete hexagonal ring, with lipids contained in the central cavity [17]. The low permeability of the urothelial barrier is believed to result from the peculiar protein array and tight junctions between umbrella cells. The urothelial barrier restricts the movement of drugs after intravesical administration and also restricts the action of the active drug fraction in the urine. Hence, many drugs fail to reach the bladder at desired therapeutic levels and ultimately lack pharmacological effects [17, 18]. Several physical approaches have been attempted in order to overcome the bladder permeability barrier in the treatment of bladder cancer and various lower urinary tract disorders. One such approach is electromotive drug administration, which combines the advantages of iontophoresis and electroporation.

Shockwave and Stone

Shockwaves are material waves, which are generated whenever different elements in a fluid approach one another, with a relative velocity higher than that of sound in the given medium. High-energy extracorporeal shockwave therapy has been used to disintegrate urolithiasis for 30 years [19]. Shockwaves for lithotripsy are generated using a self-focusing parabolic reflector mounted with piezoelectric crystals. When subjected to high voltage, compressive wave fronts are produced due to the vibration of individual crystals. Shockwave generated at the focal point of the reflector can also be harnessed for applications other than lithotripsy in lower urinary tract such as drug delivery.

Probable Mechanism of Stone Disintegration

An incoming shockwave for lithotripsy is characterized by a positive pressure pulse with a rise time of <10 ns and the negative pressure of the trailing tensile phase of the wave. The positive pressure wave and the tensile wave can create the acoustic cavitation effect, that is, the growth and collapse of microscopic bubbles [20] at the interface of the shockwave with stone. The rise of pressure in a very short time causes a very high pressure at the interface during the interaction of the shockwave and the stone. The high pressure inside the bubble and the trailing tensile phase of the shockwave can trigger extremely rapid bubble growth, followed hundreds of microseconds later by a violent collapse. Because the pressure acting on the bubble from the outside is generally not homogeneous, its collapse is asymmetrical. This may also be due to the flow drag produced by a boundary close to the bubble. One part of the bubble surface accelerates inward more rapidly than the opposite side, resulting in the development of a fluid microjet into the stone along the direction of shockwave propagation [21–24]. The fluid jet at speeds exceeding 700 m/s can penetrate the bubble and could impale the stone. The collision between the inward-moving wall of each bubble and the microjet produces a secondary shockwave. These direct and indirect shockwave effects may also contribute to stone disintegration. Bubbles generally take from 50 to 100 µs to expand after the passage of shockwaves, and they collapse after approximately 200–500 µs.

Increased Membrane Permeability from Low Energy Shockwave and Needle Free Drug Delivery

Low energy shockwave (LESW) is used clinically to improve tissue regeneration at tendon-bone junction, ischemic cardiovascular disorders, and erectile dysfunction [25–27]. The direct and indirect shockwave effects described with respect to stone disintegration could also contribute to transient membrane permeabilization on application of LESW to eukaryotic cell membranes [28, 29], and facilitate the delivery of macromolecules or drugs into cells without consequent cytotoxicity. Kodama et al. [28] suggested that shockwaves could cause shear force generated by movement of the liquid relative to the cells, to temporarily affect the permeability of the plasma membrane. The impulse (defined as the integral of pressure with duration) of the shockwave might be an important factor governing the temporary permeability increase necessary for delivering macromolecules into cells [28].

Molecular dynamics simulations demonstrated that the shockwave initially hits the membrane and is followed by a nanojet produced by the collapse of the nanobubble [30] formed due to acoustic microcavitation effect. Because the pressure acting on the bubble from the outside is generally not homogeneous, its collapse is asymmetrical, resulting in the development of a fluid microjet into the cell membrane along the direction of shockwave propagation. Therefore, LESW results in reversible permeabilization of cells in targeted areas of tissue without undue toxicity, and can deliver molecules of up to 2,000,000 molecular weight into the cytoplasm of cells [29]. The ability to deliver macromolecules of large molecular weight by cell permeabilization [31] has relevance in gene therapy and protein delivery.

LESW and Cell Membrane Poration

Studies show that [30], membrane poration by shockwave is critically dependent on the collapse of nanobubble formed due to acoustic microcavitation effect. Two-dimensional pressure maps revealed that even distribution of the shock pressure along the lateral area of the membrane is disturbed by the nanobubble, which leads to the membrane poration. The size of the pore formed depends on pressure waveform, shockwave velocity, shockwave duration, and energy flux density. The positive peak pressure of shockwaves is considered to affect the material based on the plasticity nature of the material and recent study on immortalized cell line (HEK293) and tumor-derived cell line (MCF-7) [32] reported differences in membrane permeabilization of 2 cell lines. Authors used scanning electron microscopy to report that shockwaves induced transient micro- and submicrosized deformations at the cell membrane. Cell size decreased and hole-like structures were noted after shockwave exposure. Trypan blue exclusion assays indicated that cell membranes were porated by shockwave treatment but resealed after a few seconds. Deformations of the cell membrane lasted for at least 5 min, allowing their observation in fixed cells. The mechanisms for shockwave-mediated membrane permeabilization are still poorly understood.

A recent study on osteosarcoma U2OS cell line found that shockwave increased cell membrane permeability in a P2X7 receptor-dependent manner to increase the extracellular concentration of ATP [33]. Since P2X7 receptors are also expressed in urothelium [34], they are likely to key molecular determinant of increased bladder permeability following LESW.

Intravesical Botulinum Toxin A Delivery

In August 2011, onabotulinum toxin A (Botox®, BoNT-A) was approved by the Food and Drug Administration (FDA) for the treatment of neurogenic and treatment resistant idiopathic overactive bladder. The toxin of onabotulinum toxin A is composed of 2 subunits, which are the 50 kDa light chain (L) domain and the 100 kDa heavy chain. The light chain is the catalytic moiety of the toxin. Large molecular weight of onabotulinum toxin A necessitates intradetrusor injection for clinical effectiveness [35] as topical application of onabotulinum toxin A failed to reduce bladder contractility [36].

Following intradetrusor injection, heavy chain of onabotulinum toxin A binds with high affinity receptors called synaptic vesicle proteins type 2 (SV2) expressed on parasympathetic, sympathetic and sensory neuron fibers in bladder [37]. The light chain of Onabotulinum toxin A inhibits neuroexocytosis of neurotransmitters from nerve terminals, but the translocation of the light chain of onabotulinum toxin A inside the cell is dependent on the integration of the heavy chain into the membrane bilayer [38]. Onabotulinum toxin A acts by inhibiting SNARE-dependent exocytotic processes has been demonstrated to reduce the release of various neurotransmitters, including acetylcholine, norepinephrine, calcitonin gene-related peptide, from both efferent nerves and afferent nerves [39, 40]. Onabotulinum toxin A has been applied for many medical disorders that can be benefited by chemo-denervation.

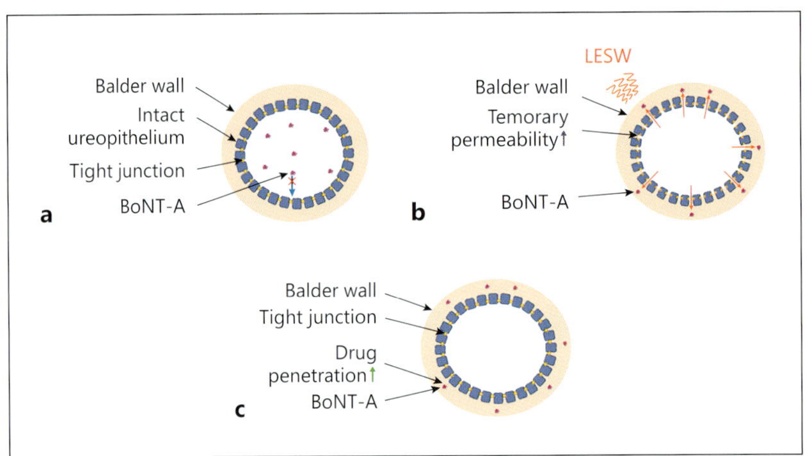

Fig. 1. a–c LESW assisted the intravesical delivery of Onabotulinumtoxin-A across urothelium by membrane poration, which was supported by significant downregulation of E-cadherin and ZO-1.

LESW and Intravesical Botulinum Toxin A Delivery

Intravesical injections of Onabotulinum toxin A have been recommended in the treatment of patients with refractory overactive bladder (OAB) and IC/PBS [40]. However, injecting substances with a needle into the bladder has some limitations, including drug leakage outside of bladder, hematuria, pain on injection sites, and uneven distribution [41, 42]. Onabotulinum toxin A with a high molecular weight of 150 kDa cannot cross the urothelium barrier and access the submucosal nerve plexus without direct injection (Fig. 1). Researchers have tried denudation of urothelium with protamine [43] or dimethyl sulphoxide DMSO [44] to assist in the delivery of Onabotulinum toxin A. Therefore, there is a need to develop a simpler and lower risk method for delivering Onabotulinum toxin-A without the need for injection.

Instead of chemically denuding urothelium, LESW represents a physical approach to cause transient reversible permeabilization of urothelium for needle free drug delivery [45]. MRI with intravesical administration of Gd-diethylenetriamine penta-acetic acid (GPA) contrast medium was used to get the visual evidence of increased rat bladder permeability after LESW (0.12 mJ/mm^2; 300 pulse) [45]. MRI established the leakage of instilled GPA outside of bladder to suggest transient increase in bladder permeability after LESW. The leakage of GPA was not observed in control animals not exposed to GPA. Although, MRI suggested increase in both mucosal (urothelium) and serosal permeability of rat bladder following LESW, but recent human study with high energy extracorporeal shockwave lithotripsy only noticed urothelial lesions and not serosal [46] to suggest that small bladder wall thickness of rat bladder is not predictive of the potential clinical toxicity from LESW.

Visual evidence of increased bladder permeability after LESW was corroborated by significant downregulation of E-cadherin (42% decrease) and ZO-1 (15% decrease). We assessed the effectiveness of LESW-assisted intravesical delivery of Onabotulinum toxin-A using the rat model of acetic acid (AA) induced bladder inflammation and bladder overactivity. Intravesical instillation of AA induced a decrease of intercontraction interval (ICI), which is considered to represent the symptom of increased urinary frequency of overactive bladder. The ICI was similarly decreased by 71.9, 72.6, and 70.6% after intravesical instillation of AA in the saline, saline plus LESW, and Onabotulinum toxin-A treated rat groups, respectively. However, the rat group that received Onabotulinum toxin-A plus LESW showed a significantly reduced response (ICI 48.6% decrease) to AA exposure without any compromise of voiding function.

The cystometric changes in the rat group treated with Onabotulinum toxin-A plus LESW were associated with a decrease in inflammatory reaction ($p < 0.05$) and expression of SNAP-23 ($p < 0.05$), SNAP25 ($p = 0.061$), and COX2 ($p < 0.05$) compared with the control group. Findings supported that LESW assisted the intravesical delivery of Onabotulinum toxin-A and in the absence of LESW, the Onabotulinum toxin-A was ineffective. It is likely that LESW potentiated the efficacy of intravesical Onabotulinum toxin-A pretreatment in suppressing the AA induced bladder hyperactivity and inflammatory reaction. Physiological and histological changes were related to the significantly reduced expression of SNAP-23, and COX2. Animal groups exposed to only Onabotulinum toxin-A or LESW failed to show any protective effect against AA-induced bladder hyperactivity and inflammation. Therefore, LESW may be a promising alternative for intravesical Onabotulinum toxin-A delivery.

Future Directions

Recently, there has been lot of interest in physical approaches to improve intravesical delivery of small molecule drugs [47] and large proteins. The transient effect of physical approaches on cell permeability is a very attractive proposition. Moreover, in certain patient populations such as pediatric populations, intradetrusor injection is not suitable and there is need for alternatives. Iontophoresis was recently used to improve Onabotulinum toxin-A delivery to children affected by neurogenic detrusor overactivity from myelomeningocele [48]. Based on our findings, LESW is a safer alternative for pediatric and elderly patient population, considering its safety.

Recent studies show that LESW can induce the expression of brain derived neurotrophic factor [49] and this effect could be valuable in used of LESW as a therapeutic approach for underactive bladder. The anti-inflammatory effect of LESW demonstrated recently [50] can be combined with drugs for synergistic effect in IC. Based on studies performed so far, bladder thickness [45, 46] is a key determinant in potential toxicity from LESW and therefore, future studies need to be carried out in animals with varying thickness of bladder.

Conclusions

Several past reports based on cell culture studies [51] have encouraged the use of LESW to assist delivery of genes, proteins and large molecular drugs. Recently, our group was the first to demonstrate the in vivo evidence of the untapped potential of LESW for intravesical drug delivery. Transient increased bladder permeability following LESW exposure has been verified by MRI and histological findings. LESW-induced transient increase in urothelial permeability is likely dependent on the acoustic microcavitation and shear stress. LESW holds immense potential in assisting the intravesical delivery of genes, proteins, and large molecular drugs.

References

1 Hsu CC, Chuang YC, Chancellor MB: Intravesical drug delivery for dysfunctional bladder. Int J Urol 2013;20:552–562.

2 Shen Z, Shen T, Wientjes MG, O'Donnell MA, Au JL: Intravesical treatments of bladder cancer: review. Pharm Res 2008;25:1500–1510.

3 Highley MS, van Oosterom AT, Maes RA, De Bruijn EA: Intravesical drug delivery. Pharmacokinetic and clinical considerations. Clin Pharmacokinet 1999;37:59–73.

4 Tyagi P, Kashyap MP, Kawamorita N, Yoshizawa T, Chancellor M, Yoshimura N: Intravesical liposome and antisense treatment for detrusor overactivity and interstitial cystitis/painful bladder syndrome. ISRN Pharmacol 2014;2014 :601653.

5 Rajaganapathy BR, Jayabalan N, Tyagi P, Kaufman J, Chancellor MB: Advances in therapeutic development for radiation cystitis. Low Urin Tract Symptoms 2014;6:1–10.

6 Kuo HC, Liu HT, Chuang YC, Birder LA, Chancellor MB: Pilot study of liposome-encapsulated onabotulinumtoxina for patients with overactive bladder: a single-center study. Eur Urol 2014;65:1117–1124.

7 Eldrup J, Thorup J, Nielsen SL, Hald T, Hainau B: Permeability and ultrastructure of human bladder epithelium. Br J Urol 1983;55:488–492.

8 Parsons CL, Boychuk D, Jones S, Hurst R, Callahan H: Bladder surface glycosaminoglycans: an epithelial permeability barrier. J Urol 1990;143:139–142.

9 Kaufman J, Tyagi V, Anthony M, Chancellor MB, Tyagi P: State of the art in intravesical therapy for lower urinary tract symptoms. Rev Urol 2010;12:e181–e189.

10 GuhaSarkar S, Banerjee R: Intravesical drug delivery: challenges, current status, opportunities and novel strategies. J Control Release 2010;148:147–159.

11 Parkin J, Shea C, Sant GR: Intravesical dimethyl sulfoxide (DMSO) for interstitial cystitis – a practical approach. Urology 1997;49(5A suppl):105–107.

12 Kaufman J, Hensley H, Jacobs J, Anthony M, Tyagi V, Tyagi P, et al: Non-invasive imaging of near infrared dye labeled liposomes facilitates evaluation of bioresidence time. J Urol 2010;183:e628.

13 Melicow MM: The urothelium: a battleground for oncogenesis. J Urol 1978;120:43–47.

14 Apodaca G: The uroepithelium: not just a passive barrier. Traffic 2004;5:117–128.

15 Lu Z, Yeh TK, Tsai M, Au JL, Wientjes MG: Paclitaxel-loaded gelatin nanoparticles for intravesical bladder cancer therapy. Clin Cancer Res 2004;10:7677–7684.

16 Khandelwal P, Abraham SN, Apodaca G: Cell biology and physiology of the uroepithelium. Am J Physiol Renal Physiol 2009;297:F1477–F1501.

17 Min G, Zhou G, Schapira M, Sun TT, Kong XP: Structural basis of urothelial permeability barrier function as revealed by Cryo-EM studies of the 16 nm uroplakin particle. J Cell Sci 2003;116:4087–4094.

18 Tammela T, Wein AJ, Monson FC, Levin RM: Urothelial permeability of the isolated whole bladder. Neurourol Urodyn 1993;12:39–47.

19 Chaussy C, Schmiedt E, Jocham D, Brendel W, Forssmann B, Walther V: First clinical experience with extracorporeally induced destruction of kidney stones by shock waves. J Urol 1982;127:417–420.

20 Bekeredjian R, Bohris C, Hansen A, Katus HA, Kuecherer HF, Hardt SE: Impact of microbubbles on shock wave-mediated DNA uptake in cells in vitro. Ultrasound Med Biol 2007;33:743–750.

21 Junge L, Ohl CD, Wolfrum B, Arora M, Ikink R: Cell detachment method using shock-wave-induced cavitation. Ultrasound Med Biol 2003;29:1769–1776.

22 Ohl CD, Ikink R: Shock-wave-induced jetting of micron-size bubbles. Phys Rev Lett 2003;90:214502.

23 Ohl CD, Wolfrum B: Detachment and sonoporation of adherent HeLa-cells by shock wave-induced cavitation. Biochim Biophys Acta 2003;1624:131–138.

24 Ohl SW, Klaseboer E, Szeri AJ, Khoo BC: Lithotripter shock wave interaction with a bubble near various biomaterials. Phys Med Biol 2016;61:7031–7053.

25 Wang CJ: An overview of shock wave therapy in musculoskeletal disorders. Chang Gung Med J 2003;26:220–232.

26 Di Meglio F, Nurzynska D, Castaldo C: Cardiac shock wave therapy: assessment of safety and new insights into mechanisms of tissue regeneration. J Cell Mol Med 2012;16:936–942.

27 Vardi Y, Appel B, Kilchevsky A, Gruenwald I: Does low intensity extracorporeal shock wave therapy have a physiological effect on erectile function? Short-term results of a randomized, double-blind, sham controlled study. J Urol 2012;187:1769–1775.

28 Kodama T, Hamblin MR, Doukas AG: Cytoplasmic molecular delivery with shock waves: importance of impulse. Biophys J 2000;79:1821–1832.

29 Kodama T, Doukas AG, Hamblin MR: Shock wave-mediated molecular delivery into cells. Biochim Biophys Acta 2002;1542:186–194.

30 Adhikari U, Goliaei A, Berkowitz ML: Mechanism of membrane poration by shock wave induced nanobubble collapse: a molecular dynamics study. J Phys Chem B 2015;119:6225–6234.

31 Lauer U, Bürgelt E, Squire Z, Messmer K, Hofschneider PH, Gregor M: Shock wave permeabilization as a new gene transfer method. Gene Ther 1997;4:710–715.

32 López-Marín LM, Millán-Chiu BE, Castaño-González K, Aceves C, Fernández F, Varela-Echavarría A, et al: Shock wave-induced damage and poration in eukaryotic cell membranes. J Membr Biol 2017;250:41–52.

33 Qi B, Yu T, Wang C, Wang T, Yao J, Zhang X, et al: Shock wave-induced ATP release from osteosarcoma U2OS cells promotes cellular uptake and cytotoxicity of methotrexate. J Exp Clin Cancer 2016;Res35:161.

34 Negoro H, Urban-Maldonado M, Liou LS, Spray DC, Thi MM, Suadicani SO: Pannexin 1 channels play essential roles in urothelial mechanotransduction and intercellular signaling. PLoS One 2014;9:e106269.

35 Smith CP, Franks ME, McNeil BK, Ghosh R, de Groat WC, Chancellor MB, et al: Effect of botulinum toxin A on the autonomic nervous system of the rat lower urinary tract. J Urol 2003;169:1896–1900.

36 Howles S, Curry J, McKay I, Reynard J, Brading AF, Apostolidis A: lack of effectiveness of botulinum neurotoxin A on isolated detrusor strips and whole bladders from mice and guinea-pigs in vitro. BJU Int 2009;104:1524–1529; discussion 1529–1530.

37 Giannantoni A, Proietti S, Vianello A, Amantini C, Santoni G, Porena M: Assessment of botulinum A toxin high affinity SV2 receptors on normal human urothelial cells. J Urol 2011;185:e319.

38 Fu FN, Busath DD, Singh BR: Spectroscopic analysis of low pH and lipid-induced structural changes in type A botulinum neurotoxin relevant to membrane channel formation and translocation. Biophys Chem 2002;99:17–29.

39 Smith CP, Chancellor MB: Emerging role of botulinum toxin in the treatment of voiding dysfunction. J Urol 2004;171:2128–2137.

40 Jiang YH, Liao CH, Kuo HC: Current and potential urological applications of botulinum toxin A. Nat Rev Urol 2015;12:519–533.

41 Boy S, Schmid M, Reitz A, Von Hessling A, Hodler J, Schurch B: Botulinum toxin injections into the bladder wall – a morphological evaluation of the injection technique using magnetic resonance imaging. J Urol 2006;5:299.

42 Kuo HC: Urodynamic evidence of effectiveness of botulinum A toxin injection in treatment of detrusor overactivity refractory to anticholinergic agents. Urology 2004;63:868–872.

43 Vemulakonda, VM, Somogyi GT, Kiss S, Salas NA, Boone TB, Smith CP: Inhibitory effect of intravesically applied botulinum toxin A in chronic bladder inflammation. J Urol 2005;173:621–624.

44 Shimizu S, Wheeler M, Saito M, Weiss R, Hittelman A: Effect of intravesical botulinum toxin A delivery (using DMSO) in rat overactive bladder model. J Urol 2012;187:e370.

45 Chuang YC, Huang TL, Tyagi P, Huang CC: Urodynamic and immunohistochemical evaluation of intravesical botulinum toxin A delivery using low energy shock wave. J Urol 2016;196:599–608.

46 Mustafa M, Aburas H, Helo FM, Qarawi L: Electromagnetic and electrohydraulic shock wave lithotripsy-induced urothelial damage: is there a difference? J Endourol 2017;31:180–184.

47 Riedl CR, Stephen RL, Daha LK, Knoll M, Plas E, Pflüger H: Electromotive administration of intravesical bethanechol and the clinical impact on acontractile detrusor management: introduction of a new test. J Urol 2000;164:2108–2111.

48 Ahmadi H, Montaser-Kouhsari L, Kajbafzadeh AM: Intravesical electromotive botulinum toxin type A administration: preliminary findings for the treatment of children with myelomeningocele and refractory neurogenic detrusor overactivity. Eur Urol Suppl 2010;9:217.

49 Wang B, Ning H, Reed-Maldonado AB, Zhou J, Ruan Y, Zhou T: Low-intensity extracorporeal shock wave therapy enhances brain-derived neurotrophic factor expression through PERK/ATF4 signaling pathway. Int J Mol Sci 2017;18:pii:E433.

50 Wang HJ, Lee WC, Tyagi P, Huang CC, Chuang YC: Effects of low energy shock wave therapy on inflammatory moleculars, bladder pain, and bladder function in a rat cystitis model. Neurourol Urodyn 2017; 36:1440–1447.

51 Ha CH, Lee SC, Kim S, Chung J, Bae H, Kwon K: Novel mechanism of gene transfection by low-energy shock wave. Sci Rep 2015;5:12843.

Yao-Chi Chuang, MD
Center for Shockwave Medicine and Tissue Engineering
Kaohsiung Chang Gung Memorial Hospital, Chang Gung University College of Medicine
123 Ta Pei Road, Niao Song Hsiang, Kaohsiung Hsien 83301 (Taiwan)
E-Mail chuang82@ms26.hinet.net

Tyagi · Chuang

Wang C-J, Schaden W, Ko J-Y (eds): Shockwave Medicine.
Transl Res Biomed. Basel, Karger, 2018, vol 6, pp 127–139 (DOI: 10.1159/000485070)

Application of Extracorporeal Shockwave Therapy on Erectile Dysfunction and Lower Urinary Tract Inflammatory Diseases

Hung-Jen Wang[a, b]

[a]Center for Shockwave Medicine and Tissue Engineering and [b]Department of Urology, Kaohsiung Chang Gung Memorial Hospital, Chang Gung University College of Medicine, Kaohsiung, Taiwan

Abstract

High-energy extracorporeal shockwave therapy has been successfully used to disintegrate urolithiasis for 30 years. However, at lower energy levels, shockwaves have enhanced expression of vascular endothelial growth factor, endothelial nitric oxide synthase, proliferating cell nuclear antigen, chemoattractant factors, and recruitment of progenitor cells for tissue regeneration and induce neovascularization. It has also improved inflammatory condition and repair injured nerve. Low energy shockwave (LESW) therapy has been used clinically with musculoskeletal disorders, ischemic cardiovascular disorders. Furthermore, LESW has been proposed to use in the urology field including erectile dysfunction, nonbacterial chronic prostatitis, and chronic pelvic pain syndromes. This chapter describes the biological effect of LESW and its current application and potential development in functional and pathological disorders in urology.

© 2018 S. Karger AG, Basel

Introduction

High-energy extracorporeal shockwave (SW) therapy has been used for the treatment of urolithiasis since1982 [1]. However, if we lower the energy SWs might enhance expression of vascular endothelial growth factor (VEGF), endothelial nitric oxide synthase (eNOS), proliferating cell nuclear antigen, and recruitment of progenitor cells, and tissue regeneration [2]. In 1987, a focused extracorporeal SW therapy was reported to have potential therapeutic benefits with delay and nonunion fractures [3–5]. In recent decades, many new applications for extracorporeal shockwaves treatment (ESWT) have been developed including heel spurs, Achilles tendons, spastic movement disorders, cellulite, wound healing, and thrombolysis treatment [6–11]. Currently, ESWT is applied in a wide range of different disorders in orthopedics (tendinopathies, cartilage repair, and bone healing); dermatology (wound healing, fire burns, ulcers, and diabetic foot ulcers); neurology (spasticity or spastic hypertonia, and neuron protection) [12–21]. ESWT may also be a promising method for use in functional urology.

Biological Effects of LESW

The biological effect of low energy SWs (LESW) has been observed through the modulation of various mechanisms, depending on different tissue types and conditions. These effects include the induction of cell proliferation and stem cell recruitment, angiogenesis, nerve regeneration, and anti-inflammation.

Cell Proliferation

Shockwaves enhances the expression of VEGF, eNOS, proliferating cell nuclear antigen, transforming growth factor-β1 (TGF-β1), fibroblast growth factor-2, chemoattractant factors, and recruitment of progenitor cells, and improves tissue regeneration at low energy level. LESW therapy facilitates tissue regeneration through cell recruitment and proliferation as first step. Severe molecular pathways (BMPs, WNTs, TGF-β1, VEGF, Ras, Erk1/2, p38 MAPK pathways) have been involved in the recruitment of specific cell, such as tendon fibroblasts, chondrocytes, human bone marrow stromal cells, primary osteoblasts, adipose tissue-derived mesenchymal stem cells, and human Jurkat T cell into target site [19, 20, 22–25]. In a streptozotocin-induced diabetic bladder dysfunction model, LESW treatment on the urinary bladder showed recruiting EdU + CD34– endogenous stem cells and the cells release abundant nerve growth factor (NGF) and VEGF. The voiding dysfunction ameliorated by LESW could be medicated by NGF and VEGF secreting [26].

Angiogenesis

LESW also induces angiogenesis. The effect of angiogenesis was mediated by increased expression of the NO and VEGF [27]. Wang et al. [27] applied LESW into the tendon-bone junction in rabbits. The shockwave induces the expression of angiogenesis-related factors, such as eNOS and VEGF, and induces neovessel growth and tissue regeneration. In a porcine ischemia heart model, LESW has also shown its ability to improve myocardial function and blood flow by increasing capillary density and VEGF expression [28]. The mechanism of angiogenesis has been suggested to be through mechanosensory complex formation, including VEGF2, vascular endothelial cadherin, platelet endothelial cell adhesion molecule 1, and stimulates phosphorylation of Akt, eNOS and angiogenic gene expression [29]. In patients with coronary artery disease receiving Li-ESWT, tissue regeneration in patients might be associated with angiogenesis-related cytokines such as eNOS, basic fibroblast growth factor (bFGF), stromal cell-derived factor 1, and its receptor C-X-C motif chemokine 4 (CXCR4) [30, 31].

Nerve Regeneration

SW has pain relief effects after application to musculoskeletal systems. Previous studies have postulated that ESWT was observed to have immediate analgesic effect. It could be the effect of degeneration of intercutaneous nerve fibers. However, this raised a question of safety in ESWT, as some studies observed damage to the myelin sheath after ESWT [32–34]. Hausdorf et al. [33] applied moderate energy (1,500 pulses, energy flux density 0.9 mJ/mm^2) on the hind limbs of rabbits; this resulted in a substantial loss of unmyelinated nerve within the femoral nerve of treated the limb [33]. Another study demonstrated an effect, positively correlated with the intensity of the SWs (2,000 pulses with 0.08, 0.19, and 0.49 mJ/mm^2), of inducing reversible segmental demyelination in large nerve fibers in rats [34]. However, as observed previously, different energy intensities of ESWT induce different biological effects. At a lower energy intensity (300 pulses, 0.1 mJ/mm^2), ESWT was able to induce a growth-promoting effect in myelinated motor axonal regeneration in autograft sciatic nerve graft rat models [35]. Other studies applied low energy ESWT to injured spinal cords of Sprague-Dawley rats [36, 37]. CD31 and α-SMA expression was promoted, and TUNEL-positive cells were decreased in the injured spinal cord in ESWT group [37].

Schwann cells dysfunction in the peripheral nervous system was related to some traumatic injury or diabetic diseases. Recent innovations in tissue engineering may be able to differentiate different stem cell populations into Schwann cells. These cells can be transplanted into patients as a potential cell replacement therapy for peripheral neuropathic conditions [38]. Li-ESWT has been proven to improve native Schwann cell isolation and culture in vitro without phenotype commitment and reducing senescence-associated markers after 15 passages [39]. Another study demonstrated that p75 gene and p-Erk1/2, which are regarded as Schwann cell activation-related markers, were upregulated after Li-ESWT had a positive effective on Schwann cell proliferation in vitro [40]. A combination of Li-ESWT and tissue engineering technologies may be potential technology achieving cell requirements for clinical therapeutic application.

Anti-Inflammation

ESWT also triggers anti-inflammatory reaction. This action has been shown to be related to the mechanism of "mechanotherapy" that induces different biological reactions' pathways in immunomodulation [41]. In some preclinical studies, LESW significantly decreased infiltration of inflammatory cells such as neutrophils and macrophages in burned skin [42] or skin flaps [43], ischemic heart muscle [44, 45], and in the urinary bladder [46, 47]. This effect should be associated with the suppression of pro-inflammatory cytokines, chemokines, and MMPs. Pro-inflammatory cytokines

(IL-1α, IL-4, IL-6, IL-10, IL-12, IL-13 and INF-γ) were significantly suppressed after ESWT in murine models. Chemokines (CCL2, CCL3, CCL4, CCL7, CXCL1, CXCL2, CXCL5) and matrix metallopeptidase (MMP-3, MMP-9, MMP-13) were also reduced after ESWT treatment in murine burned skin, skin isografts, and urinary bladders [42, 46, 48]. In the modulation of macrophages function, Li-ESWT not only decreased the infiltration but also down regulated M1 marker genes (CD80, COX-2, CCL5) expression in M1 macrophages, and promote a synergistic effect on M2 marker genes (ALOX15, MRC1, CCL18) expression in M2 macrophages [49]. This leads to a decreased pro-inflammatory profile in M1 macrophages and an increase in the presence of M2 macrophages with an anti-inflammatory profile [44, 49, 50].

Application of Extracorporeal Shockwave Therapy in Genital Urinary Diseases

Erectile Dysfunction
Erectile dysfunction is a common disease primarily affecting men older than 40 years. The incidence of this disorder increases with age. A report from International Consultation Committee for Sex Medicine showed that the prevalence of erectile dysfunction range from 2–9% in men between 40–49 years old to 20–40% in men between 60–69 years old and even 50–100% in men over 70 years old [51]. By the year of 2025, the prevalence of erectile dysfunction in the world has been predicted to reach 322 million [52]. So it has been regarded as a major health problem in the male aging group. According to several epidemiological studies, erectile dysfunction has association with cardiovascular disease. Because the arteries size of penis is smaller than coronary artery, the same level of endothelial dysfunction causes a more significant reduction of blood flow in erectile tissue and lead to dysfunction [53]. Treatment for erectile dysfunction is including lifestyle modification, vacuum constrictive devices, intracavernosal injection, transurethral therapy, penile prostheses, and phosphodiesterase type 5 inhibitor (PDE5i) [54, 55]. Treatment strategies for erectile dysfunction have continued to focus on restoration of natural erection. PDE5i was introduced 2 decades ago, which is now regarded as the first-line treatment for erectile dysfunction. However, there is a substantial erectile dysfunction population (range 30–40%) that is not satisfied with PDE5i treatment, including patients taking nitrate medications, those who intolerant to the side effect of PDE5i, or those with certain types of ED that are refractory to PDE5i treatment [56]. PDE5i should be used with caution in severe or uncontrolled hypertension patients, unstable angina, or patients taking a-blocker for blood pressure control [54]. As we mentioned above, there is some association between erectile dysfunction and cardiovascular disease. A novel treatment should be innovated for this group of patients.

ESWT has biological effects on angiogenesis, cell proliferation, and NO synthesis, which might have positive therapeutic effects on endothelial dysfunction and vasculogenic erectile dysfunction. In 2008, Muller et al. [57] evaluated the impact of SW on functional and structural changes in erectile tissue in rats. They used radial SWs applied to the penis at varied energy and dose levels, from 1 bar/1,000 SW/1 session to 2 bar/2,000 SW/3 sessions. In this study both low and high energy levels of shockwave resulted in collagenization of corporal smooth muscle, with increased apoptosis in the erectile tissue [57]. This raised the issue about whether corporal tissue and erectile function would be impaired by shockwave therapy. However, the following studies using focused ESWT with energy levels from 0.06 mJ/mm^2 300 pulses to 0.09 mJ/mm^2 1,000 pulses on the penis showed different results [40, 58]. In a diabetes mellitus (DM)-associated ED rat model, they injected streptozotocin into new born male rats. After DM was induced, 300 shock at energy level 0.1 mJ/mm^2 and 120 min frequency was applied to the penis of these DM rats ($n = 8$). Streptozotocin injection induced a significant decrease in erectile function which might be due to the decrease in endothelial and smooth muscle contents and the number of nNOS-positive nerves. Endogenous mesenchymal stem cells (5-ethynyl-2-deoxyuridin positive cells) were significantly higher in numbers in DM + SW than in DM rats. Li-ESWT can partially ameliorate DM-related ED by promoting regeneration of nNOS-positive nerves, endothelium, and smooth muscle in the penis [58]. Low energy ESWT-induced progenitor cell recruitment and Schwann cell proliferation also led to nerve regeneration and angiogenesis in a pelvic neurovascular injured rat model [40].

The clinical application of LESW on the treatment of erectile dysfunction was initially observed on patients who received ESWT for Peyronie's disease. ESWT has been used in the treatment of Peyronie's disease since 1989. The results have been favorable in relieving pain, with overall improvement in pain score reported to be 51.6% (66 of 128) and 18.8% (12 of 64) of patients experienced relief and complete remission of pain in ESWT and control group, respectively. However, regarding sex function, the results were inconsistent. Several studies showed improvement in erectile function using IIEF-ED dominate score evaluation (IIEF score increased from 14 to 19.4 at 24 weeks) [59, 60]. The other studies showed that ESWT had no therapeutic effect on Peyronie's disease-related erectile dysfunction [61, 62]. In a meta-analysis study of extracorporeal SW therapy for Peyronie's disease, 296 out of 443 patients with PD had complained of varying degrees of ED. The recovery rate of patients in the ESWT group was 39.9% (55 of 138) compared with 29.7% in the control group. However, the difference was not significant (OR 2.22, 95% CI 0.69–7.11, $p = 0.18$) [63].

Dr. Vardi et al. [64] first reported that LESW had a positive clinical effect in restoring erectile function in a randomized controlled trial (RCT) in patients who have previously responded to oral PDE5i. The SWs were delivered equally into 5 locations in genital area including distal, mid and proximal penile shaft, and the left, right crura. Each location received 300 shocks at 0.09 mJ/mm^2 with frequency of 120/min, which ended up with 1,500 shocks per section. The 9-week treatment protocol comprised of 2 treatment sessions per week for 3 weeks that were repeated after a 3-week no-treatment interval. The International Index of Erectile Function-Erectile Function domain score (IIEF-EF) in the treated group increased by 6.7 points, which was a significant

Table 1. Erectile dysfunction: basic characteristics and results of clinical studies

Reference	Year	No of patients (ESWT/ sham)	Type of ED	ESWT Device	Protocol	IIEF-EF change (treatment/ sham)	*p* value of IIEF-EF change after ESWT	Study type
Vardi et al. [64]	2012	40/20	PDE5i responders	Omnispec ED1000, Medispec Ltd	2 sessions 1,500 SW, Efd 0.09 mJ/ mm^2 in every week for 3 weeks, repeated after 3 weeks' break, total 18,000 SW	6.7/3.0	0.032	RCT
Clavijo et al. [66]	2017	379/223, from 7 RCT studies	PDE5i responders	Omnispec ED1000, Medispec Ltd (6 studies) Piezo Wave 2, Richard Wolf GmbH (1 study)	2 sessions 1,500 SW, Efd 0.09 mJ/ mm^2 in every week for 3 weeks, repeated after 3 weeks' break, total 18,000 SW 1 treatment per week/10 weeks/ total 6,000 SW (PiezoWave 2)	6.40 vs. 1.65	<0.0001 (baseline to follow-up) 0.047 (between treatment and sham)	Meta-analysis
Kitrey et al. [68]	2016	37/18	PDE5i Non-responders	Omnispec ED1000, Medispec Ltd.	2 sessions 1,500 SW, Efd 0.09 mJ/ mm^2 in every week for 3 weeks, repeated after 3 weeks' break, total 18,000 SW	5/0	<0.005	RCT

SW, shockwave; IIEF-EF, International Index of Erectile Function-Erectile Function domain; Efd, Energy Flux Density.

difference from the sham group (3 points, $p = 0.0322$). No side effect such as pain or ecchymosis was reported in the series. Currently, there are 3 major SW machines used in the field for ED treatment.

A meta-analysis of 14 studies including 833 patients with ED, receiving ESWT from 2005 to 2015 [65], showed that LESW significantly improved the IIEF score and Erection hardness score (EHS; risk difference 0.16, $p < 0.01$) in ED patients without Peyronie's disease for at least 3 months (mean difference 2.36; $p < 0.0001$). Therapeutic efficacy was better in mild to moderate ED patients, or in the ESWT combined with PDE5i usage group, regardless of disease severity. Another meta-analysis focused on 7 RCTs, involving 602 PDE5i responders with an average follow-up of 19.8 months. Pooled change in the IIEF-EF score, from baseline to follow-up after LESW, improved significantly compared with the sham group (6.4 points; 95% CI 1.78–11.02; $p < 0.001$ vs. 1.65 points; 95% CI 0.92–2.39; $p < 0.0001$; between-group difference $p = 0.047$) [66]. Recent studies regarding ESWT on PDE5i responders were summarized in Table 1.

For patients with severe ED who are nonresponsive to PDE5i, LESW has been shown to be able to allow these patients to become responders [67]. In a prospective double-blind RCT in patients with vasculogenic ED who were not taking PDE5i due to lack of effect, 54% of those receiving LESW in combination with PDE5i achieved hard enough erections, in comparison with 0% of the sham group (PDE5i only). The IIEF-EF score increased from 7 to 13 at follow-up in the LESW group, compared with the incremental improvement (from 8 to 8.5) in the sham group [68]. This indicated that even with severe ED, patients were non-responders to PDE5i. SW therapy was able to allow nearly half of them to achieve an erection sufficiently rigid for completing sexual intercourse.

Inflammatory Disorders: Non-Infectious Prostatitis and Chronic Pelvic Pain

Nonbacterial prostatitis and chronic pelvic pain syndrome (CPPS) are common types of prostatitis, affecting 5–10% of the male population. It is a frustrating disease defined as genital urinary pain with or without voiding symptoms in the absence of uropthogenic bacteria or malignancy. A possible cause of nonbacterial prostatitis may be neurogenic or immunogenic inflammation. This inflammatory process activates prostate-afferent nerves, and induces prostate pain and referred pain. The imbalance of the altered androgen receptor hormone condition is one of the factors inducing prostatodynia. An imbalance of cytokines including INF-γ, IL-6, IL-10, TNF-α, and IL-1β also induces an autoimmune process [69]. Furthermore, the prostate may not be the source of pelvic pain, which neurotrophin NGF may implicate in neurogenic pain and inflammation. Previous studies have demonstrated that NGF and cytokines (IL-6 and IL-10) that regulate inflammation may play a role in the pain symptoms experienced by patients with CPPS [70]. Changes of NGF level in expressed prostatic secretions occur in proportion to pain severity in expressed prostatic secretions [71]. Many cytokines and growth factors, including TNF-α, IL1-β, TGF-α and TGF-β1 can upregulate NGF production [72]. Finally, psychological stress may have an impact on measurable biochemical change and influence the above-mentioned factors [69]. So far there are several treatment options for nonbacterial prostatitis; however, the clinical efficacy is limited [73–75]. It is imperative to develop a novel therapy for patients with refractory chronic prostatitis and prostatic pain. Previous studies have demonstrated that LESW has protective effects on inflammatory reaction, by lowering the expression of NGF, IL-6, IL-12, TNF-α, COX-2 and iNOS [46, 47]. These effects may be used as a novel therapy in treating nonbacterial prostatitis/CPPS.

The first randomized double-blind placebo controlled study about the effects of ESWT on CPPS was conducted by an Austria group [76]. Patients with type IIIb prostatitis (CPPS) received one perineally applied ESWT weekly for 4 weeks (maximal energy flow density 0.25 mJ/mm^2, 3,000 shocks, Duolith SD1, Storz Medical). The SW transducer position changed every 500 shocks to scan virtually the entire prostate and pelvic floor area. In the treatment group, NIH-chronic prostatitis symptoms index (NIH-CPSI) improved by 16.7% continuously to 12 weeks and visual analog score (VAS) improved 50% after 4 and 12 weeks. At 12 weeks, the ESWT group achieved 25% improve in IPSS and 10.5% in IIEF. Compared with the control group, ESWT significantly improved pain and micturition, as well as erectile function and quality of life [77]. Dr. Al Edwan et al. [78] further demonstrated the maximal effects of LESW (ESWT once a week for one month, with a protocol of 2,500 pulses on perineal with total energy flow density 0.25 mJ/mm^2 over 13 min) on the IPSS, NIH-CPSI, and the American Urological Association Quality of Life due to Urinary Symptoms scale (AUA QoL US) at 2 weeks; the effect was stabilized until 12 months . Current clinical studies on ESWT for CPPS/chronic pelvic pain were summarized in Table 2 [77–81]. There were no significant side effects mentioned in these

Table 2. Chronic pelvic pain syndrome: basic characteristics and results of clinical studies

Reference	Year	No of patients (ESWT/sham)	Type of disease	ESWT device	Protocol	NIH-CPSI mean difference (ESWT/sham)	p value	Study type
Zimmermann et al. [77]	2009	30/30	Type IIIB prostatitis	Duolith SD1 device (Storz Medical)	1 treatment per week/ 4 weeks/3,000 pulses, total Efd: 0.25 mJ/mm²	–16.7 vs. 2.1% (4 weeks after treatment)	<0.001 vs. 0.865	RCT
Zeng et al. [81]	2012	40/40	Type IIIB prostatitis/ CPPS	HB-ESWT 01 (Haibin medica, China)	5 treatment per week/ 2 weeks/2,000 pulses, total Efd: 0.06 mJ/mm²	71.1 vs. 27% patient improved (at endpoint of study)	<0.001	RCT
Vahdatpour et al. [80]	2013	20/20	Type IIIB prostatitis/ CPPS	Duolith SD1 device (Storz Medical)	1 treatment per week/ 4 weeks/3,000 pulses, total Efd: 0.25 mJ/mm², add 0.05 mJ/mm² per week	–38.5 vs. –17.3% (4 weeks after treatment)	<0.0001	RCT
Moayednia et al. [79]	2014	19/18	Type IIIB prostatitis/ CPPS	Duolith SD1 device (Storz Medical)	1 treatment per week/ 4 weeks/3,000 pulses, total Efd: 0.25 mJ/mm², add 0.05 mJ/mm² per week	No difference between 2 group at 24 weeks	ns	RCT
Al Edwan et al. [78]	2017	41	Type IIIB prostatitis/ CPPS	Electropneumatic shockwave unit (E-S.W. T Roland, pagani, Italy)	1 treatment per week/ 4 weeks/2,500 pulses, total Efd: 0.25 mJ/mm²	–9.1 (2 weeks after treatment)	0.000	Cohort

SW, shockwave; NIH-CPSI, National Institutes of Health Chronic Prostatitis Symptom Index; Efd, Energy Flux Density.

studies. However, more trials with different protocols to maximize the short and long-term outcome are needed before LESW can be widely applied in nonbacterial prostatitis/CPPS.

Inflammatory Disorders in Urinary Bladder

Interstitial cystitis/painful bladder syndrome (IC/BPS) is recognized as a chronic inflammatory disorder of the urinary bladder that manifests as storage voiding symptoms (frequency, urgency, and nocturia) with or without pain. Several possible etiologies have been proposed for IC/BPS, but currently there is no effective long-term treatment for this disorder [82]. Previous studies found that serum C-reactive protein, NGF, and other inflammatory cytokines levels increased in IC/BPS patients, suggesting that chronic inflammation plays an important role in the pathophysiology of IC/BPS [83–85]. The anti-inflammatory and immune modulation effect of ESWT made it as a potential candidate for relieving inflammatory condition of this disease.

In our preclinical studies, in a cyclophosphamide (CYP)-induced rat cystitis model, LESW also showed its potential for relieving inflammatory conditions and overactivity [47]. We directly applied 300 shocks of Li-ESWT on the urinary bladder at an energy level of 0.12 mJ/mm². LESW reduced pain behavior in conjunction with the downregulation of IL-6 (40.9%) and NGF (33%) expression at day 4 (both decline $p < 0.05$), as compared with the control. CYP-induced COX-2 expression

(38.6%, $p < 0.05$) and bladder overactivity was reduced via LESW (77.8% increase in inter-contraction interval compare with cystitis group, $p < 0.05$) at day 8. Using different doses of LESW and CYP, Dr. Chen et al. [46] evaluates the effect of Li-ESWT on bladder inflammation in vitro and in vivo. Smooth muscle cell cline (A7R5) was pretreated with Li-ESWT (200 impulses 0.12 mJ/mm^2), then cultured in high glucose medium with menadione (an oxidative stimulator) and lipopolysaccharide (LPS, an inflammatory stimulator). MMP-9, TNF-α, NF-κB, IL-1β, 4 indicators of inflammatory biomarkers and NOX-1, NOX-2, 2 indices of oxidative stress, did not differ between control group and control + Li-ESW-treated group. These suggest that Li-ESWT did not elicit the inflammatory reaction. In menadione and the LPS-treated group, these markers were higher compare with control group and were significantly reversed by Li-ESWT. For in vivo study, a total of 400 impulses of SWs at 0.11 mJ/mm^2 were delivered to urinary bladder (divided 2 sessions, 3 and 24 h after CYP injection). Li-ESWT effectively attenuated proteinuria, hematuria and lowered the IL-6 level in the urine. In bladder tissue CD74+, CD68+, COX-2, substance P and MIF positive cells were lower in the Li-ESWT group, which suggest that ESWT inhibited inflammatory cell infiltration in the bladder [46]. Therefore, we suggest that LESW may be a potential option for the treatment of bladder inflammatory conditions and overactivity.

Conclusions

LESW therapy has shown its biological tissue effects including cell proliferation and stem cell recruitment, neovascularization, anti-inflammation, neurological tissue repair. Currently, LESW therapy has been clinically used in the field of urology for the treatment of (1) lower genital urinary tract inflammatory disorders, such as non-infectious cystitis or prostatitis, through the mechanisms of anti-inflammation; (2) erectile dysfunction, through the mechanisms of neovascularization, cell proliferation, and stem cell recruitment. This novel approach, using an existing technology, shows a safe and noninvasive method and promising results for the application in urological disorders. However, the biological effect of different shockwave machines and proper protocol (energy level, impulses and treatment interval) are still not clear at the clinical level. This induced some inconsistent results in studies using different treatment protocols. More advanced basic science studies and large-scale randomized clinical trials are warranted before LESW can be further widely used.

Disclosure Statement

The author declares that there are no conflicts of interest to disclose.

References

1 Chaussy C, Schuller J, Schmiedt E, Brandl H, Jocham D, Liedl B: Extracorporeal shock-wave lithotripsy (ESWL) for treatment of urolithiasis. Urology 1984; 23:59–66.

2 Wang CJ: An overview of shock wave therapy in musculoskeletal disorders. Chang Gung Med J 2003; 26:220–232.

3 Haupt G, Haupt A, Ekkernkamp A, Gerety B, Chvapil M: Influence of shock waves on fracture healing. Urology 1992;39:529–532.

4 Stranne SK, Callaghan JJ, Fyda TM, Fulghum CS, Glisson RR, Weinerth JL, Seaber AV: The effect of extracorporeal shock wave lithotripsy on the prosthesis interface in cementless arthroplasty. Evaluation in a rabbit model. J Arthroplasty 1992;7:173–179.

5 Johannes EJ, Kaulesar Sukul DM, Matura E: High-energy shock waves for the treatment of nonunions: an experiment on dogs. J Surg Res 1994;57:246–252.

6 Cosentino R, Falsetti P, Manca S, De Stefano R, Frati E, Frediani B, Baldi F, Selvi E, Marcolongo R: Efficacy of extracorporeal shock wave treatment in calcaneal enthesophytosis. Ann Rheum Dis 2001;60: 1064–1067.

7 Perlick L, Schiffmann R, Kraft CN, Wallny T, Diedrich O: [extracorporal shock wave treatment of the achilles tendinitis: experimental and preliminary clinical results]. Z Orthop Ihre Grenzgeb 2002;140: 275–280.

8 Lohse-Busch H, Kraemer M, Reime U: [A pilot investigation into the effects of extracorporeal shock waves on muscular dysfunction in children with spastic movement disorders]. Schmerz 1997;11:108–112.

9 Siems W, Grune T, Voss P, Brenke R: Anti-fibrosclerotic effects of shock wave therapy in lipedema and cellulite. Biofactors 2005;24:275–282.

10 Schaden W, Thiele R, Kolpl C, Pusch M, Nissan A, Attinger CE, Maniscalco-Theberge ME, Peoples GE, Elster EA, Stojadinovic A: Shock wave therapy for acute and chronic soft tissue wounds: a feasibility study. J Surg Res 2007;143:1–12.

11 Rosenschein U, Yakubov SJ, Guberinich D, Bach DS, Sonda PL, Abrams GD, Topol EJ: Shock-wave thrombus ablation, a new method for noninvasive mechanical thrombolysis. Am J Cardiol 1992;70: 1358–1361.

12 van der Worp H, van den Akker-Scheek I, van Schie H, Zwerver J: ESWT for tendinopathy: technology and clinical implications. Knee Surg Sports Traumatol Arthrosc 2013;21:1451–1458.

13 Wang CJ: Extracorporeal shockwave therapy in musculoskeletal disorders. J Orthop Surg Res 2012;7: 11.

14 Dymarek R, Halski T, Ptaszkowski K, Slupska L, Rosinczuk J, Taradaj J: Extracorporeal shock wave therapy as an adjunct wound treatment: a systematic review of the literature. Ostomy Wound Manage 2014;60:26–39.

15 Butterworth PA, Walsh TP, Pennisi YD, Chesne AD, Schmitz C, Nancarrow SA: The effectiveness of extracorporeal shock wave therapy for the treatment of lower limb ulceration: a systematic review. J Foot Ankle Res 2015;8:3.

16 Lee JY, Kim SN, Lee IS, Jung H, Lee KS, Koh SE: Effects of extracorporeal shock wave therapy on spasticity in patients after brain injury: a meta-analysis. J Phys Ther Sci 2014;26:1641–1647.

17 Lobenwein D, Tepekoylu C, Kozaryn R, Pechriggl EJ, Bitsche M, Graber M, Fritsch H, Semsroth S, Stefanova N, Paulus P, Czerny M, Grimm M, Holfeld J: Shock wave treatment protects from neuronal degeneration via a toll-like receptor 3 dependent mechanism: implications of a first-ever causal treatment for ischemic spinal cord injury. J Am Heart Assoc 2015;4:e002440.

18 Venkatesh Prabhuji ML, Khaleelahmed S, Vasudevalu S, Vinodhini K: Extracorporeal shock wave therapy in periodontics: a new paradigm. J Indian Soc Periodontol 2014;18:412–415.

19 Cheng JH, Wang CJ: Biological mechanism of shockwave in bone. Int J Surg 2015;24:143–146.

20 Wang CJ, Cheng JH, Kuo YR, Schaden W, Mittermayr R: Extracorporeal shockwave therapy in diabetic foot ulcers. Int J Surg 2015;24:207–209.

21 Chen YL, Chen KH, Yin TC, Huang TH, Yuen CM, Chung SY, Sung PH, Tong MS, Chen CH, Chang HW, Lin KC, Ko SF, Yip HK: Extracorporeal shock wave therapy effectively prevented diabetic neuropathy. Am J Transl Res 2015;7:2543–2560.

22 Wang CJ: An overview of shock wave therapy in musculoskeletal disorders. Chang Gung Med J 2003; 26:220–232.

23 Yu T, Junger WG, Yuan C, Jin A, Zhao Y, Zheng X, Zeng Y, Liu J: Shockwaves increase T-cell proliferation and IL-2 expression through atp release, P2X7 receptors, and fak activation. Am J Physiol Cell Physiol 2010;298:C457–C464.

24 Wang FS, Wang CJ, Huang HJ, Chung H, Chen RF, Yang KD: Physical shock wave mediates membrane hyperpolarization and ras activation for osteogenesis in human bone marrow stromal cells. Biochem Biophys Res Commun 2001;287:648–655.

25 Raabe O, Shell K, Goessl A, Crispens C, Delhasse Y, Eva A, Scheiner-Bobis G, Wenisch S, Arnhold S: Effect of extracorporeal shock wave on proliferation and differentiation of equine adipose tissue-derived mesenchymal stem cells in vitro. Am J Stem Cells 2013;2:62–73.

26 Jin Y, Xu L, Zhao Y, Wang M, Jin X, Zhang H: Endogenous stem cells were recruited by defocused low-energy shock wave in treating diabetic bladder dysfunction. Stem Cell Rev 2017;13:287–298.

27 Wang CJ, Wang FS, Yang KD, Weng LH, Hsu CC, Huang CS, Yang LC: Shock wave therapy induces neovascularization at the tendon-bone junction. A study in rabbits. J Orthop Res 2003;21:984–989.

28 Nishida T, Shimokawa H, Oi K, Tatewaki H, Uwatoku T, Abe K, Matsumoto Y, Kajihara N, Eto M, Matsuda T, Yasui H, Takeshita A, Sunagawa K: Extracorporeal cardiac shock wave therapy markedly ameliorates ischemia-induced myocardial dysfunction in pigs in vivo. Circulation 2004;110:3055–3061.

29 Ha CH, Kim S, Chung J, An SH, Kwon K: Extracorporeal shock wave stimulates expression of the angiogenic genes via mechanosensory complex in endothelial cells: mimetic effect of fluid shear stress in endothelial cells. Int J Cardiol 2013;168:4168–4177.

30 Ping Y, Tao G, Yun-zhu P, Yu W, Hong-yan C: Gw24-e0232 effects of cardiac shock wave therapy on the angiogenesis related cytokine of cad patients. Heart 2013;99:A155–A156.

31 Fu M, Sun CK, Lin YC, Wang CJ, Wu CJ, Ko SF, Chua S, Sheu JJ, Chiang CH, Shao PL, Leu S, Yip HK: Extracorporeal shock wave therapy reverses ischemia-related left ventricular dysfunction and remodeling: molecular-cellular and functional assessment. PLoS One 2011;6:e24342.

32 Wang CJ, Huang HY, Yang K, Wang FS, Wong M: Pathomechanism of shock wave injuries on femoral artery, vein and nerve. An experimental study in dogs. Injury 2002;33:439–446.

33 Hausdorf J, Lemmens MA, Heck KD, Grolms N, Korr H, Kertschanska S, Steinbusch HW, Schmitz C, Maier M: Selective loss of unmyelinated nerve fibers after extracorporeal shockwave application to the musculoskeletal system. Neuroscience 2008;155:138–144.

34 Wu YH, Liang HW, Chen WS, Lai JS, Luh JJ, Chong FC: Electrophysiological and functional effects of shock waves on the sciatic nerve of rats. Ultrasound Med Biol 2008;34:1688–1696.

35 Hausner T, Pajer K, Halat G, Hopf R, Schmidhammer R, Redl H, Nogradi A: Improved rate of peripheral nerve regeneration induced by extracorporeal shock wave treatment in the rat. Exp Neurol 2012;236:363–370.

36 Yamaya S, Ozawa H, Kanno H, Kishimoto KN, Sekiguchi A, Tateda S, Yahata K, Ito K, Shimokawa H, Itoi E: Low-energy extracorporeal shock wave therapy promotes vascular endothelial growth factor expression and improves locomotor recovery after spinal cord injury. J Neurosurg 2014;121:1514–1525.

37 Yahata K, Kanno H, Ozawa H, Yamaya S, Tateda S, Ito K, Shimokawa H, Itoi E: Low-energy extracorporeal shock wave therapy for promotion of vascular endothelial growth factor expression and angiogenesis and improvement of locomotor and sensory functions after spinal cord injury. J Neurosurg Spine 2016;25:745–755.

38 Lehmann HC, Hoke A: Use of engineered Schwann cells in peripheral neuropathy: Hopes and hazards. Brain Res 2016;1638:97–104.

39 Schuh CM, Hercher D, Stainer M, Hopf R, Teuschl AH, Schmidhammer R, Redl H: Extracorporeal shockwave treatment: a novel tool to improve schwann cell isolation and culture. Cytotherapy 2016;18:760–770.

40 Li H, Matheu MP, Sun F, Wang L, Sanford MT, Ning H, Banie L, Lee YC, Xin Z, Guo Y, Lin G, Lue TF: Low-energy shock wave therapy ameliorates erectile dysfunction in a pelvic neurovascular injuries rat model. J Sex Med 2016;13:22–32.

41 D'Agostino MC, Craig K, Tibalt E, Respizzi S: Shock wave as biological therapeutic tool: from mechanical stimulation to recovery and healing, through mechanotransduction. Int J Surg 2015;24:147–153.

42 Davis TA, Stojadinovic A, Anam K, Amare M, Naik S, Peoples GE, Tadaki D, Elster EA: Extracorporeal shock wave therapy suppresses the early proinflammatory immune response to a severe cutaneous burn injury. Int Wound J 2009;6:11–21.

43 Kuo YR, Wang CT, Wang FS, Yang KD, Chiang YC, Wang CJ: Extracorporeal shock wave treatment modulates skin fibroblast recruitment and leukocyte infiltration for enhancing extended skin-flap survival. Wound Repair Regen 2009;17:80–87.

44 Abe Y, Ito K, Hao K, Shindo T, Ogata T, Kagaya Y, Kurosawa R, Nishimiya K, Satoh K, Miyata S, Kawakami K, Shimokawa H: Extracorporeal low-energy shock-wave therapy exerts anti-inflammatory effects in a rat model of acute myocardial infarction. Circ J 2014;78:2915–2925.

45 Lei PP, Tao SM, Shuai Q, Bao YX, Wang SW, Qu YQ, Wang DH: Extracorporeal cardiac shock wave therapy ameliorates myocardial fibrosis by decreasing the amount of fibrocytes after acute myocardial infarction in pigs. Coron Artery Dis 2013;24:509–515.

46 Chen YT, Yang CC, Sun CK, Chiang HJ, Chen YL, Sung PH, Zhen YY, Huang TH, Chang CL, Chen HH, Chang HW, Yip HK: Extracorporeal shock wave therapy ameliorates cyclophosphamide-induced rat acute interstitial cystitis though inhibiting inflammation and oxidative stress-in vitro and in vivo experiment studies. Am J Transl Res 2014;6:631–648.

47 Wang HJ, Lee WC, Tyagi P, Huang CC, Chuang YC: Effects of low energy shock wave therapy on inflammatory moleculars, bladder pain, and bladder function in a rat cystitis model. Neurourol Urodyn 2017; 36:1440–1447.

48 Stojadinovic A, Elster EA, Anam K, Tadaki D, Amare M, Zins S, Davis TA: Angiogenic response to extracorporeal shock wave treatment in murine skin isografts. Angiogenesis 2008;11:369–380.

49 Sukubo NG, Tibalt E, Respizzi S, Locati M, d'Agostino MC: Effect of shock waves on macrophages: a possible role in tissue regeneration and remodeling. Int J Surg 2015;24:124–130.

50 Tepekoylu C, Lobenwein D, Urbschat A, Graber M, Pechriggl EJ, Fritsch H, Paulus P, Grimm M, Holfeld J: Shock wave treatment after hindlimb ischemia results in increased perfusion and M2 macrophage presence. J Tissue Eng Regen Med 2016, [Epub ahead of print].

51 Nicolosi A, Moreira ED Jr, Shirai M, Bin Mohd Tambi MI, Glasser DB: Epidemiology of erectile dysfunction in four countries: cross-national study of the prevalence and correlates of erectile dysfunction. Urology 2003;61:201–206.

52 Ayta IA, McKinlay JB, Krane RJ: The likely worldwide increase in erectile dysfunction between 1995 and 2025 and some possible policy consequences. BJU Int 1999;84:50–56.

53 Gandaglia G, Briganti A, Jackson G, Kloner RA, Montorsi F, Montorsi P, Vlachopoulos C: A systematic review of the association between erectile dysfunction and cardiovascular disease. Eur Urol 2014; 65:968–978.

54 Decaluwe K, Pauwels B, Boydens C, Van de Voorde J: Treatment of erectile dysfunction: new targets and strategies from recent research. Pharmacol Biochem Behav 2014;121:146–157.

55 Shamloul R, Ghanem H: Erectile dysfunction. Lancet 2013;381:153–165.

56 Dorsey P, Keel C, Klavens M, Hellstrom WJ: Phosphodiesterase type 5 (PED5) inhibitors for the treatment of erectile dysfunction. Expert Opin Pharmacother 2010;11:1109–1122.

57 Muller A, Akin-Olugbade Y, Deveci S, Donohue JF, Tal R, Kobylarz KA, Palese M, Mulhall JP: The impact of shock wave therapy at varied energy and dose levels on functional and structural changes in erectile tissue. Eur Urol 2008;53:635–642.

58 Qiu X, Lin G, Xin Z, Ferretti L, Zhang H, Lue TF, Lin CS: Effects of low-energy shockwave therapy on the erectile function and tissue of a diabetic rat model. J Sex Med 2013;10:738–746.

59 Palmieri A, Imbimbo C, Creta M, Verze P, Fusco F, Mirone V: Tadalafil once daily and extracorporeal shock wave therapy in the management of patients with peyronie's disease and erectile dysfunction: results from a prospective randomized trial. Int J Androl 2012;35:190–195.

60 Skolarikos A, Alargof E, Rigas A, Deliveliotis C, Konstantinidis E: Shockwave therapy as first-line treatment for peyronie's disease: a prospective study. J Endourol 2005;19:11–14.

61 Chitale S, Morsey M, Swift L, Sethia K: Limited shock wave therapy vs. sham treatment in men with peyronie's disease: results of a prospective randomized controlled double-blind trial. BJU Int 2010;106: 1352–1356.

62 Poulakis V, Skriapas K, de Vries R, Dillenburg W, Ferakis N, Witzsch U, Melekos M, Becht E: Extracorporeal shockwave therapy for Peyronie's disease: an alternative treatment? Asian J Androl 2006;8:361–366.

63 Gao L, Qian S, Tang Z, Li J, Yuan J: A meta-analysis of extracorporeal shock wave therapy for peyronie's disease. Int J Impot Res 2016;28:161–166.

64 Vardi Y, Appel B, Kilchevsky A, Gruenwald I: Does low intensity extracorporeal shock wave therapy have a physiological effect on erectile function? Short-term results of a randomized, double-blind, sham controlled study. J Urol 2012;187:1769–1775.

65 Lu Z, Lin G, Reed-Maldonado A, Wang C, Lee YC, Lue TF: Low-intensity extracorporeal shock wave treatment improves erectile function: a systematic review and meta-analysis. Eur Urol 2017;71:223–233.

66 Clavijo RI, Kohn TP, Kohn JR, Ramasamy R: Effects of low-intensity extracorporeal shockwave therapy on erectile dysfunction: a systematic review and meta-analysis. J Sex Med 2017;14:27–35.

67 Gruenwald I, Appel B, Vardi Y: Low-intensity extracorporeal shock wave therapy – a novel effective treatment for erectile dysfunction in severe ed patients who respond poorly to PDE5 inhibitor therapy. J Sex Med 2012;9:259–264.

68 Kitrey ND, Gruenwald I, Appel B, Shechter A, Massarwa O, Vardi Y: Penile low intensity shock wave treatment is able to shift pde5i nonresponders to responders: a double-blind, sham controlled study. J Urol 2016;195:1550–1555.

69 Pontari MA, Ruggieri MR: Mechanisms in prostatitis/chronic pelvic pain syndrome. J Urol 2008; 179:S61–S67.

70 Miller LJ, Fischer KA, Goralnick SJ, Litt M, Burleson JA, Albertsen P, Kreutzer DL: Nerve growth factor and chronic prostatitis/chronic pelvic pain syndrome. Urology 2002;59:603–608.

71 Watanabe T, Inoue M, Sasaki K, Araki M, Uehara S, Monden K, Saika T, Nasu Y, Kumon H, Chancellor MB: Nerve growth factor level in the prostatic fluid of patients with chronic prostatitis/chronic pelvic pain syndrome is correlated with symptom severity and response to treatment. BJU Int 2011;108:248–251.

72 Yoshida K, Gage FH: Cooperative regulation of nerve growth factor synthesis and secretion in fibroblasts and astrocytes by fibroblast growth factor and other cytokines. Brain Res 1992;569:14–25.

73 Rees J, Abrahams M, Doble A, Cooper A, Prostatitis Expert Reference Group (PERG): Diagnosis and treatment of chronic bacterial prostatitis and chronic prostatitis/chronic pelvic pain syndrome: a consensus guideline. BJU Int 2015;116:509–525.

74 Parker J, Buga S, Sarria JE, Spiess PE: Advancements in the management of urologic chronic pelvic pain: what is new and what do we know? Curr Urol Rep 2010;11:286–291.

75 Cohen JM, Fagin AP, Hariton E, Niska JR, Pierce MW, Kuriyama A, Whelan JS, Jackson JL, Dimitrakoff JD: Therapeutic intervention for chronic prostatitis/chronic pelvic pain syndrome (CP/CPPS): a systematic review and meta-analysis. PLoS One 2012;7:e41941.

76 Zimmermann R, Cumpanas A, Hoeltl L, Janetschek G, Stenzl A, Miclea F: Extracorporeal shock-wave therapy for treating chronic pelvic pain syndrome: a feasibility study and the first clinical results. BJU Int 2008;102:976–980.

77 Zimmermann R, Cumpanas A, Miclea F, Janetschek G: Extracorporeal shock wave therapy for the treatment of chronic pelvic pain syndrome in males: a randomised, double-blind, placebo-controlled study. Eur Urol 2009;56:418–424.

78 Al Edwan GM, Muheilan MM, Atta ON: Long term efficacy of extracorporeal shock wave therapy (EWST) for treatment of refractory chronic abacterial prostatitis. Ann Med Surg (Lond) 2017;14:12–17.

79 Moayednia A, Haghdani S, Khosrawi S, Yousefi E, Vahdatpour B: Long-term effect of extracorporeal shock wave therapy on the treatment of chronic pelvic pain syndrome due to non bacterial prostatitis. J Res Med Sci 2014;19:293–296.

80 Vahdatpour B, Alizadeh F, Moayednia A, Emadi M, Khorami MH, Haghdani S: Efficacy of extracorporeal shock wave therapy for the treatment of chronic pelvic pain syndrome: a randomized, controlled trial. ISRN Urol 2013;2013:972601.

81 Zeng XY, Liang C, Ye ZQ: Extracorporeal shock wave treatment for non-inflammatory chronic pelvic pain syndrome: a prospective, randomized and sham-controlled study. Chin Med J (Engl) 2012;125:114–118.

82 Dyer AJ, Twiss CO: Painful bladder syndrome: an update and review of current management strategies. Curr Urol Rep 2014;15:384.

83 Chung SD, Liu HT, Lin H, Kuo HC: Elevation of serum C-reactive protein in patients with OAB and IC/BPS implies chronic inflammation in the urinary bladder. Neurourol Urodyn 2011;30:417–420.

84 Liu HT, Kuo HC: Increased urine and serum nerve growth factor levels in interstitial cystitis suggest chronic inflammation is involved in the pathogenesis of disease. PLoS One 2012;7:e44687.

85 Tyagi P, Barclay D, Zamora R, Yoshimura N, Peters K, Vodovotz Y, Chancellor M: Urine cytokines suggest an inflammatory response in the overactive bladder: a pilot study. Int Urol Nephrol 2010;42:629–635.

Hung-Jen Wang, MD
Center for Shockwave Medicine and Tissue Engineering, Department of Urology
Kaohsiung Chang Gung Memorial Hospital, Chang Gung University College of Medicine
No.123, Dapi Rd., Niaosong Dist., Kaohsiung 83301 (Taiwan)
E-Mail hujewang@yahoo.com.tw

Section V: Future Prospects of Shockwave Medicine

Wang C-J, Schaden W, Ko J-Y (eds): Shockwave Medicine.
Transl Res Biomed. Basel, Karger, 2018, vol 6, pp 140–157 (DOI: 10.1159/000485072)

Current Applications and Future Prospects of Extracorporeal Shockwave Therapy

Valerio Sansone[a, b] · Roberto Frairia[c] · Manuel Brañes[d] ·
Pietro Romeo[a] · Maria Graziella Catalano[c] · Rachel C. Applefield[a]

[a]IRCCS Galeazzi Orthopaedic Institute, and [b]Department of Biotechnology and Biosciences, University of
Milano-Bicocca, Milan, and [c]Department of Medical Sciences. University of Torino, School of Medicine, Torino,
Italy; [d]Faculty of Medicine, Institute of Biomedical Sciences, University of Chile, Santiago, Chile

Abstract

The scope of medical applications for extracorporeal shockwaves (ESWs) continues to widen as our understanding of the mechanisms behind their diverse functionalities increases. The first successful application of ESWs was in the treatment of kidney stones via shockwave lithotripsy. Subsequently, new uses for ESWs, as shown in the other sections of this book, were proposed and have since become viable treatments. The impressive spectrum of indications for shockwaves (SWs), ranging from ischemic cardiac disease to musculoskeletal disorders among others, demonstrates their great potential for benefitting patients presenting disparate diagnoses. On the horizon, there are sure to be more discoveries for new medical applications for SWs. In this chapter, we present some of the newer and lesser known medical applications for shockwaves: mesenchymal stem cells, cancer therapy, wound healing, and biofilm treatment. In addition, it is well known that SWs are effective in treating certain medical disorders as a standalone treatment; in this chapter, however, we present various treatment schemes that show how SWs can be used in combination with other treatment therapies to achieve even better outcomes, opening the door to more effective methods of curing some of the most treatment-resistant pathologies in medicine. © 2018 S. Karger AG, Basel

Orthopedic Applications for Bone Regeneration Strategies with Mesenchymal Stem Cells: A Possible Role for Extracorporeal Shockwaves

Several orthopedic conditions require the availability of a large amount of bone stock, such as skeletal reconstruction of large bone defects (caused by trauma, infection, tumor resection), fracture nonunions, skeletal abnormalities, or when the regenerative process is impaired, such as in the case of ischemic tissues and osteoporosis. Often the request is far beyond the normal potential of self-healing, and thus there is an irrefutable need to develop innovative therapies to enhance the bone healing course. Under normal circumstances, bone homeostasis is based on a tightly coupled balance between anabolism (bone formation) and catabolism (bone resorption). When this balance is lost, bone tissue reacts thanks to an innate capacity of self-repair. This reaction, that is bone healing, requires an intricate coordination of various cellular and mechanosensitive processes which involves cells, extracellular matrices, and signaling molecules [1]. However, the mechanism can be impaired for several reasons, resulting in delayed bone healing or nonunion. In these cases, surgery and/or bone regeneration strategies are required for boosting fracture healing. Several therapeutic options are available, either alone or in combination. Traditional therapies encompass less invasive bone stimulation techniques through more invasive interventions [2–5]. The success of each of these options is however variable with inconsistent results [6]. Nevertheless, the current gold standard remains biological autologous bone grafting which has a limited supply, unpredictable reparative potential and is associated with morbidities related to the harvesting procedure [7]. Another recent option is osteobiologics, a term that covers all the class of engineered materials created for promoting bone healing [8]. The list of products is rapidly expanding due to the increasing request for grafts in orthopedics, maxillofacial surgery, and dentistry. Even though the use of allografts has been successful, the demand has grown much faster than the number of donors, spurring the development of artificial materials with the osteoinductive and osteoconductive properties of the natural bone graft. A further step consisted in adding osteogenic cells and growth factors. However, the value of current products is uncertain because of the lack of adequate clinical studies necessary to establish their usefulness.

Basic Science Principles in Bone Defect Healing: The Role of Mesenchymal Stem Cells
Mesenchymal stem cells (MSCs) are cells that are able to (1) differentiate into various mesenchymal lineages, including osteoblasts and (2) regenerate the tissue of origin in vivo, while also contributing to its maintenance and repair, thus showing multipotency and self-renewal capacity [9]. Bone marrow is the most traditional source of MSCs, although they have been isolated from almost all body tissues. MSCs are likely based in the perivascular environment of all vascularized tissues, being pericytes and adventitial cells of blood vessels practically indistinguishable from MSCs [10, 11]. They are crucial for bone homeostasis as bones undergo continuous remodeling throughout adulthood and during fracture healing. Bone remodeling is based on bone resorption carried out by osteoclasts, followed by the recruitment of osteoblasts, cells of mesenchymal origin that replace the erased bone. An exact coupling between resorption and replacement is essential for successful bone remodeling and both cell types need to be constantly refilled by tissue-specific stem cells. The signals that regulate the decision of progenitor cells to turn into osteoblasts are complex and only partially known [12]. Transforming growth factor-β triggers the activation of osteoblast

specific genes and Runx2 is considered the major transcription factor to give rise to the activation of osteoblast specific genes [13].

The other condition in which MSCs play a substantial regenerative role is in response to injury. Cells can differentiate and release several cytokines and growth factors with immunomodulatory trophic and chemotactic effect on injured tissues [14]. Trophic action relies on the production of substances inhibiting apoptosis and fibrosis and on the stimulation of angiogenesis through the release of vascular endothelial growth factor (VEGF) [15]. Immunomodulatory effects are mediated by direct cell-cell contact and through the production of bioactive substances acting on dendritic cells, B and T cells [16]. Chemotactic activity is exerted on immune cells and endothelial cell progenitors.

Basic Science Principles in Bone Defect Healing: The Role of Extracorporeal Shockwaves Treatment

The positive effects of extracorporeal shockwaves treatment (ESWT) on fracture healing in animal experiments are well known [17–19]. The most relevant findings of many studies are that shockwaves (SW) promotes the growth of MCSs and differentiation toward osteoprogenitors through the transforming growth factor-β and VEGF induction [20, 21]. Similar results have been reported in the clinical setting for ESWT in long bone fractures nonunions with a success rate ranging from 50 to 85% [22–24]. Cacchio et al. [25] in a randomized clinical trial concluded that ESWT was as effective as surgery in achieving healing of long-bone hypertrophic nonunions. However, results of ESWT in atrophic nonunions were definitively less successful. Wang et al. [18] observed 80% of hypertrophic long bone nonunions healing, whereas only 27% of atrophic non-unions were cured. Similar results were confirmed by other authors, although Xu et al. [27] observed an overall healing rate of 75.4% with none of the atrophic nonunions healed [26]. In this context, the question about the real effectiveness of ESWT in these conditions arises. A positive answer comes from the studies on neovascularization induced by ESWT in treated tissues, given that adequate vascularity is essential for stem cell migration and survival [28, 29]. Through the upregulation and expression of several angiogenic and osteogenic growth factors, bone healing can be significantly stimulated [30–32]. A clinical study on a series of femoral atrophic nonunions confirmed these observations. Lesions were treated with endomedullary fixation followed by ESWT, with an overall success rate of 63.6%, whereas the success rate was 100% when ESWT was applied within 12 months of surgery [33]. Nevertheless, the most interesting result concerned patients who did not heal and requested a second surgical treatment of bone grafting and plating augmentation: all of them were healed within 5 months. These data support the hypothesis of a possible facilitating effect of the ESWT preceding surgery in terms of tissue preconditioning. A comparable scenario, although in a different clinical setting, is the successful preventive use of SW in the prophylaxis of wound healing disturbances after vein harvesting for coronary artery by-pass graft surgery [34].

Bone Nonunions as an Example of New Potential Use of ESWT

As stated above, there is a significant clinical need to develop innovative approaches for the treatment of fracture nonunions and large bone defects. MCSs-based therapy may be a promising tool in the field of regenerative medicine, thanks to several advantageous features of MSCs as they are (1) multipotent cells that can migrate to injury sites; (2) capable of suppressing the local immune response; and (3) available in abundant quantities from the patients themselves [35]. Furthermore, MSCs show plasticity and ability to respond to external physical stimuli with the enhanced growth of bone-marrow osteoprogenitor cells. On the other hand, shockwaves were revealed to be able to promote biological processes in MSCs, including increased proliferation, survival, and migration [36, 37]. Furthermore, ESWT stimulates bone healing through the upregulation of bone morphogenetic proteins, enhances trabecular bone volume and thickness of the treated bone, and induces osteogenic differentiation of adipose-derived SCs into osteoblast-like cells [38–41]. Consequently, ESWT may represent a potential tool to manipulate MSC behavior for clinical applications.

Atrophic nonunion is associated with a deficiency of MSCs at the fracture site; thus, the hypothesis to stimulate resident MSCs with the aim of boosting tissue healing seems adequately justified [28]. This approach appears to have great potential as it suggests that mechanical stress preconditions MSCs for enhanced therapeutic performance without any genetic manipulation or loss of differentiation capacity and senescence observed with cell culture serial passaging [42]. Consequently, a new perspective in the use of ESW for ameliorating the outcome of the usual ESW treatment of bone nonunions might consist of a combined treatment:

1. SW stimulation of selected bone donor sites (i.e., femoral greater trochanter area) of the same patient. The aim would be to enhance stem and progenitor cell recruitment triggering resident MSCs to replicate and differentiate toward osteogenic lineage.

2. After an adequate time span (i.e., 4 weeks) a bone marrow aspirate can be carried out from treated zones, followed by the concentration and consequent percutaneous reimplantation of autologous MSCs in tissue-defective areas.

3. Concomitant ESW treatment of the target site, with the aim of laying the groundwork for transplanted cells, inducing local neovascularization and trophic changes.

The advantages of a model based on percutaneous injection are obvious, as this delivery method would substantially reduce morbidity over open surgical implant techniques. The procedure might be clearly repeated if necessary, and could be used for all the other clinical conditions in which implemented bone regeneration is requested.

In conclusion, the use of ESWT for expanding MSCs multipotent differentiation capacity and their equally central role as cellular modulators might have the potential to open a new horizon for the treatment of nonunions and bone defects. ESW stimulations are a dynamic approach to manipulate human MSC behavior with the aim of exploiting their full regenerative capacity. The same preconditioning model may be also applied in other mesenchymal tissues for different therapeutic needs.

Extracorporeal Shockwave: A Real Prospect in Malignant Tumor Treatment

Introduction

A crucial prerequisite for any successful cancer therapy is that the benefits should outweigh deleterious side effects; taking this into account, it is admitted that any chemotherapy scheme can be limited by both suboptimal specificity for cancer cells and the probability to induce suppression of the host anticancer immunity. More potent drugs have been continually developed to increase the potency of the therapies. As the molecular power of chemotherapeutic agents increased, unfortunately, the toxicity and undesired side effects increased as well. Among the promising research areas, investigating new and better therapeutic management of cancer, such as immunotherapy, gene therapy, and the development of new pharmaceutical schemes, the search for innovative methods to deliver drugs is another plausible strategy to help reduce chemotherapy toxicity. Recently, extracorporeal shockwaves (ESW) have been proposed as a method for enhanced drug delivery.

The interaction of mechanical stress with biological tissue has become a recurrent topic. With great experimental progress and the rapid developments in computer hardware systems such as high-precision hydrophones, which allow for a very accurate characterization of the shockwave profiles, we are able today to interpret the basic effects of ESW at the cellular and sub-cellular levels [43]. Such interpretations permit the translated use of shockwaves as a new target for drug delivery.

The Impact of Shockwaves on Cells

Shockwave therapy has been applied to cells and soft tissues in different clinical situations, including attempts to treat cancer, often with inconclusive success [44, 45]. At present, we are limited by a basic understanding of the relationship between ESW and tumors. In vitro studies have shown that ESW lithotripsy can cause various types of cell damage ascribed both to the positive and negative pressure phase as well as cavitation [46–51]. Nevertheless, little is known with regard to the way in which cavitation bubbles develop within tissues. ESW's ability to exert extremely high-pressure gradients on cell membranes had induced the belief that they could achieve a cytotoxic effect as additional support in cancer treatments [52, 53]. However, more recent in vitro and in vivo studies show that SW could only elicit a temporary growth delay, despite the considerable morphological changes observed at the cellular level in the plasma membrane, mitochondria, cytoplasm, and nucleus [54–56]. The permeabilization of the cell membrane seems to be the most prominent alteration induced in cancer cells by sublethal doses of high energy ESW, thereby increasing the diffusion of various molecules from the surrounding medium into the cells [51–53, 55–58]. In an adherent model of cell growth on transparent substrates, fluorescent microscope analysis demonstrated that upon exposure to ESW, the cells adjacent to cavitation bubbles, subject to the protracted permeabilization of the membrane, appear completely damaged and detached from the surface, whereas the cells further away from the bubbles survived [59]. The initiation of cell damage has been attributed to the irreversible deformations of the cytoskeleton and inspired in vivo experiments based on the appliance of ESW to tumors [60]. Encouraging results were derived from the delayed tumor growth in mice, the reduction of tumor size in nude rats, and by the complete remission of dorsal skin tumors in hamsters [61–63]. It was therefore supposed that permeabilizing energies increased the mortality of neoplastic cells in a dose-dependent manner [64, 65]. Several other in vitro studies on different cell lines in culture were performed, even though because of the large heterogeneity of mechanical, biological, and analytical variables of each experiment the comparison of the different results is difficult [32, 44, 54, 66–69]. Moreover, the impact of ESW on cell survival varies not only based on the cell type but also based on identical cell lines [70]. Radial ESW reduced the relative number of HFFF2 (Human Fetal Foreskin Fibroblasts) cells in the G0/G1-phase and increased the relative number of cells in the G2/M phase of the cell cycle, confirming that the in vitro exposure of cells to ESW has a cell type specific effect on cell number and proliferative activity [71].

Cellular damage induced by ESW has been ascribed to the collision between cells or to the generation of free radicals; however, cavitation and generation of jet streams in the extracellular environment is currently the most reputed hypothesis. Survived cells are still able to form tumors; however, they are smaller than in untreated controls [54, 70, 72].

Shockwaves and Anticancer Drugs
Since it was supposed that the delivering of a physical energy into a tumor could potentiate the effect of cytotoxic molecules by increasing their uptake into the neoplastic cells, studies have been conducted on the combined treatment of suspended tumor cells with ESW and some anticancer agents. When applied in vitro, ESW induce a transient increase in the permeability of cell membranes, thereby allowing higher drug concentrations into the cell, thus eliciting a significant enhancement of drug cytotoxicity [73–76].

This methodology may be useful for some macromolecules like Type-1 ribosome-inactivating proteins or the highly cytotoxic anticancer agent saporin [57]. Previously, the same authors had shown that ESW, generated by the ablation of a target material by a short laser pulse, may be implemented to deliver fluorescent macromolecules into cells in vitro achieving the membrane permeabilization without toxic effects for the cells as shown by the MTT assay [77, 78].

A similar effect could be obtained through electroporation, a technique which employs high-voltage electric pulses and that has been shown to be able to enhance the action of bleomycin applied to subcutaneous tumor nodules [79]. It must be pointed out, nevertheless, that a major obstacle of electrochemotherapy is that electrodes must be in close contact with the whole tumor surface. This excludes its use in internal organs, whereas ESW does not require direct contact and can be focused internally in a noninvasive procedure. Cells of human estrogen-dependent breast cancer (MCF-7) have revealed to be responsive to the combined treatment with ESW and paclitaxel, an antimicrotubule agent active against a variety of solid tumors [64].

In addition, ESW may cause apoptosis, or programmed cell death, that has been attributed to the cavitation [64, 76]. Apoptosis is an important physiological process whose goal is to eliminate damaged or redundant cells during normal tissue development and that plays a key role in the surveillance against tumors [80, 81]. Apoptosis can be triggered by a variety of antitumor drugs, radiation, or electric fields characterized by pulse duration in the range of a nanosecond [82, 83]. Moreover, induction of apoptosis has been described to occur in response to a variety of stress signals, including ESW even though studies about this topic are few [76, 84–86]. Tumoral cell lysis has been recently ascribed to the intracellular DNA damage induced by high overpressure and shear forces generated by ESW [87]. The cytotoxic effect of ESW could become more effective when used in combination with chemotherapies. Indeed, combined treatment on cell lines from human osteosarcoma, human colorectal adenocarcinoma and anaplastic thyroid cancer, elicited a significant enhancement of cytotoxicity and the induction of apoptosis in cancer cells [88–90].

Sonodynamic/Photodynamic Properties of Shockwaves
As documented in human and in experimental animal models, solid tumor cells produce and accumulate porphyrin [91, 92]. The administration of the tumor-localizing photosensitizer 5-aminolevulinic acid (ALA), a precursor in the heme biosynthetic pathway, has been proposed in photodynamic therapy (PDT) because of the cytotoxic effect succeeding the activation of the photosensitizer by light [93].

Although the last 25 years have witnessed the advent of PDT as an effective anticancer treatment, the shallow penetration of the light through skin and the deep tissues limits PDT to the treatment of superficial and endoscopically reachable tumors [94]. To overcome this limitation, Catalano et al. [90] adopted an innovative "sonodynamic/photodynamic" technique based on the activation of ALA by means of ESW treatment. The "sonodynamic" procedure is based on the property of nonthermal ultrasounds to activate sonosensitizer agents such as hematoporphyrins able to kill cancer cells by generating highly reactive oxygen species [95–97]. Thus, transferring such concepts to ESW, the inertial cavitation precipitated by acoustic waves at appropriate energy flux rate may transfer a certain amount of energy to the sonosensitizer agent, triggering the excitation of the molecule's electrons [98]. The advantageous feature of this novel option for cancer treatment emerges from the possibility to focus the acoustic energy on malignancy sites lying deep within the tissues, overcoming one of the

                    Sansone · Frairia · Brañes · Romeo · Catalano · Applefield

main drawbacks of PDT. The cavitation induced by ESW results in a confined concentration of energy that is sufficient to generate the sono-luminescent emission as well as to trigger a photochemical process resulting in the formation of the cytotoxic singlet oxygen [99].

Combined ALA-shockwave treatment significantly inhibited the HT-29 (human colorectal carcinoma) cell growth [89]. Non-inertial as well as inertial cavitation induces apoptosis and inhibits cell growth by increasing the G0/G1 population through the intracellular activation of protoporphyrin IX (PPIX). This technique resulted effective in vivo in inducing necrosis and apoptosis in breast cancer and colon cancer implanted in animal samples [100, 101]. Most recently, the cytotoxic anticancer effect of sonodynamic therapy (SDT), using ESW-activated PPIX, has been established on a syngeneic rat breast cancer model by MRI. The SDT-treated group showed a significant decrease in MRI tumor size measurements 72 h after treatment with the PPIX precursor ALA and ESW [102].

MSC generate reactive oxygen species and, in turn, affect cancer cell growth when in co-culture with human glioblastoma (U87) or osteosarcoma (U2OS) cells exposed to medium-energy ESW with a significant increase in both apoptosis and necrosis of cancer cells [103]. These data positively compares with the "bilayer sonophore" theory, which is as follows: under appropriate conditions, the bilayer membrane can convert the oscillating acoustic pressure waves into nanometric and micrometric intracellular deformations, which are subsequently able to induce intracellular cavitation [104].

Nano-Scale Technology, Shockwaves, and SDT
Current advances in material science and bioengineering encourage the application of nanoscale technologies in medicine. Nanocarriers, or nanoencapsulation systems, have been extensively studied also in the context of tumor treatment. Compared with traditional dosage forms, nanocarriers possess specific characteristics in chemotherapeutic delivery, such as low toxicity, high drug-loading, controlled drug release, targeting tumor site and delivering of different types of drugs [105–107]. Many types of nanocarriers have been introduced in tumor diagnosis and treatments including inorganic nanoparticles (NPs), micelles, quantum dots, and polymers [106, 108, 109]. NPs carry the loaded drugs to the tumor site through the blood stream making use of the enhanced permeability and of the retention effect caused by the defective vascular architecture of the tumor tissue [110]. Some nanocarriers have been approved by the Food and Drug Administration for cancer therapy, and several have already been in preclinical or clinical studies [111, 112]. NPs tend to accumulate in the vasculature at the margins of the tumor and seldom inside the neoplasia [113]. Hence, the whole delivery process in vivo can be impaired by the atypical tumor vessels and the disordered interstitium.

The use of stimuli-responsive NPs offers a new opportunity for drug and gene therapy comprising the physical targeting of drugs and genes by external stimuli like magnetic fields, ultrasound, light, and heat [114]. Over the last 15 years, ultrasound energy has drawn attention to the targeted-drug by sonication after the injection of micellar encapsulated drugs [115]. In addition to targeted tumor uptake, this method allows the uniform distribution of drug-carrying micelles throughout the tumor tissue. It has been demonstrated that sonosensitizers conjugated to NPs are more readily taken up by cells with respect to the free drug [116]. Moreover, stimuli-responsive nanocarriers may enhance the SDT response after being properly engineered with a protoporphyrin derivative.

It was recently described that the combined treatment with doxorubicin-loaded glycol chitosan nanobubbles (NBs) and ESW enhances cytotoxicity of doxorubicin in anaplastic thyroid cancer cells [117, 118]. Cavitation has a double effect: it triggers the perturbation of the NBs resulting in drug release and it induces a transient permeabilization of both cell membranes increasing drug intake.

Recent data show that the combined treatment with both paclitaxel- or docetaxel-loaded NBs and ESWs greatly enhance the cytotoxicity of taxanes and reduces cancer cell migration in human castration-resistant prostate cancer cells [119]. Similar results have been observed in vivo with the use of pulsed laser [120]. It must be taken into account that NPs are not only effective as sensitizers per se but also as energy transducers [121].

An innovative sonosensitizing system was recently described by using ESW and a new formulation of NPs (poly-methyl methacrylate core-shell NPs) loaded with the porphyrin derivative TPPS (TPPS-NPs). The sonodynamic treatment TPPS-NPs/ SWs were investigated with regards to the cytotoxic effect on the human neuroblastoma cell line, SH-SY5Y [122]. The sonodynamic treatment (ESW in combination with TPPS-NPs) showed a statistically significant decrease in SH-SY5Y cell proliferation. This new sonosensitizing system significantly decreased cancer cell growth even in a three-dimensional model of neuroblastoma, postulating its potential application for furthered progress in sonodynamic anticancer therapy [122].

Cancer Gene Therapy and Shockwave-Aided Gene Transfer
Currently, more than 64% of all gene therapy trials worldwide are aimed at the treatment of solid or hematological malignancies, making cancer gene therapy a dominant area in both basic and clinical research [123].

Physical methods – such as microinjection, ballistic gene delivery, electroporation, sonoporation, and laser irradiation – were found to be effective for transfecting primary cells, progenitor cells, and stem cells through in vitro, ex vivo, and in vivo approaches [124, 125].

Enhancement of the immune response through the augmentation of natural cytokines is a promising methodology. Immunogen therapy with plasmid interleukin-12, has been tested in some immunogenic mouse models; however, the serious potential systemic toxicity should be considered [126, 127]. In order to prevent the downregulation of immunological responses, gene delivery methods, including electroporation, liposomal bubbles, NPs, recombinant adenoviruses, alphaviral vector, and acoustic cavitation have been adopted to transfer interleukin-12 gene [128–130]. Because of their adaptability to in vivo purposes, easy-to-use feature, and with minimal effects on normal physiology, electroporation and acoustic cavitation seem to be the most promising techniques for gene therapy applications [131]. The combination of ESW and DNA cationization has shown to be as effective as those reported by other acoustic cavitation-based approaches [132, 133].

ESW-aided gene transfer has been associated with many advantages in cancer gene therapy: localization of DNA transfer to the tumor; cavitation-induced cell killing resulting in tumor ablation, and the avoidance of antigenic responses associated with virus-based DNA delivery methods [134]. Recently, a synergistic approach has been described using ultrasound and NPs to deliver plasmid DNA to cancer cells, achieving a really high transfection efficiency and enhancement of the antitumor effect [135, 136]. The efficacy of ESW for permeabilizing human cells and promoting transfection with both cationic lipid-assembled and naked DNA has recently been confirmed [65]. This promising strategy to introduce genetic materials into targeted cells, at virtually any depth in a receptor-independent way shows the capacity of an in vivo plasmid DNA or RNA delivery to almost any solid organ of interest without imparting cellular or nucleic acid damage [125, 136].

Concluding Remarks

We see an innovative strategy in the treatment of cancer – a strategy that integrates with existing therapy protocols:

1. ESW acts as an "ultrasound-susceptibility modification agent," as they are able to modify the permeability of cell membranes allowing a better delivery of chemotherapeutic drugs into cytosol;

2. ESW enhances both the cytotoxic activities of photosensitizers as well as the apoptotic signal transduction pathway: they can act as a further tool in "sonodynamic/photodynamic" therapy;

3. Gene transfer can be induced by ESW treatment in vivo, particularly with enhanced acoustic cavitation, which supports the concept that "Gene and shockwave therapy might be advantageously merged."

Peripheral Nervous System Lesions and Spinal Cord Injuries

Considering the long recovery and rehabilitation process after injuries of peripheral nerves, there is a great need for the development of procedures that promote peripheral nerve regeneration and thereby reduce the related health-care and social costs. Current knowledge about the effects of ESWT on nervous system is incapacitated by limitations, as the research is restricted to animal and in vitro studies. Nerve injury activates molecular mechanisms and signaling pathways, which lead to the de-differentiation of quiescent highly specialized myelinating and non-myelinating Schwann cells (SCs) as well as their reprogramming into proliferative progenitors [137]. SCs orchestrate the repair with a double role: the removal of degenerated tissue and the bringing in of the re-myelination of newly generated fibers [138]. When traumatic nerve interruption occurs, primary reconnection for the appropriate guidance of nerve regrowth is indicated. In larger nerve defects, autograft is the gold standard, but it could result ineffective if scarring and sprouting occur [139]. Nerve regeneration is slow and sometimes incomplete and unpredictable similar to functional recovery [140]. In this respect, the support given by ESWT may be strategic both in promoting nerve regeneration and in avoiding sequelae as fibrous substitution or development of neuromas. In sciatic nerve autograft, low energy ESWT induced the early enhancement of the functional recovery rate [141]. At the same time, the number of advanced myelinated fibers and regenerate axons, which has been related to macrophage infiltration, was higher in the ESWT group. These results have been interpreted as a clue for a faster Wallerian degeneration with the removal of deteriorated axons and preparation for the following nerve regeneration. A positive reaction of SCs to ESWT has been observed in vitro with a high proliferation rate and higher expression, compared to untreated cells of specific markers of myelinating cells and of regenerative phenotypes [142]. Effective results after low-energy ESWT have also been observed in experimental spinal cord injuries (SCI), with reduced neuronal loss and consistent improvement of motor functions, without any significant neural tissue damage [143]. Enhancement of neural angiogenic sites and a lesser degree of apoptosis have also been reported along with a significant increase of the neurotrophic factors together with macrophages and SCs activity [144, 145].

VEGF is demonstrated to be a potent neurotrophic factor, which has a protective effect on injured neurons, and it promotes the survival of the neurons exerting an antiapoptotic effect, encouraging the axonal regeneration and improving functional recovery [146, 147]. VEGF also plays a significant role in spinal cord repair and, afterwards, the activation of VEGF and its receptors by ESWT may restore blood supply.

An auxiliary role of ESWT to cell therapy has been evaluated in experimental models of spinal cord injury. Cell transplantation of induced pluripotent stem cells with intravenous delivery is less successful than direct intralesional grafting [148]. To improve the results of intravenous delivery, SWs have been used in a model of chronic SCI with preconditioning the microenvironment of the injured tissues before treatment: the number of engrafted MSCs resulted higher with an increased expression of neurotrophic growth factors [149].

Another relevant consequence of nerve injury is muscle atrophy due to denervation. In an experimental sciatic nerve damage, ESWT was effective in counteracting the changes induced by the inhibition of muscle contraction and the decrease of protein synthesis. The level of muscular atrophy was reduced and functional recovery improved [150].

In conclusion, original experiences highlight the valuable role of ESW in the treatment of peripheral nervous system damages. The potential use of ESWT after nerve injury may be:

1. Inducing a faster and more effective Wallerian degeneration to facilitate nerve regrowth after direct suture or grafting.

2. Providing aid to intravenous delivery of MSCs-cell therapy in case of spinal cord injury.

3. Hindering distal amiotrophy after peripheral nerve injury.

4. Increasing the VEGF levels with neurotrophic effects.

Moreover, recent literature data suggest that the nervous system, like musculoskeletal tissues, might have multipotent perivascular, mechanosensitive resident stem cells that contribute to neurogenesis and vasculogenesis [151]. Working on the assumption that SWs would induce neoangiogenesis and exert a positive effect on MSCs, it seems sensible to hypothesize an analogue role on nervous tissues, legitimating the prospective use of ESW for activating resident stem cells in their native environment. However, the distinctive features of the nervous cells and tissue would require specific studies aiming to establish the most effective and safest energy levels.

Implant-Related Infections in Orthopedics: ESWT as an Antibiofilm Agent

Orthopedic implant-related infections are a major cause of joint arthroplasty and osteosynthesis failure. Biofilms occur when bacteria colonize synthetic implants in vivo and form an extracellular polymeric slime matrix. Because of this protective barrier, antibiotics are much less effective in treating biofilms as opposed to other types of bacterial infections. In recent years, various antibiofilm agents have been proposed, although their effective role in the clinical setting remains to be established.

Often the only treatment for a biofilm is to surgically remove and replace the implant; however, these revisions are costly and undesirable for the patient. In an in vitro study of plastic and metal implants, Kizhner et al. [152] showed that laser-generated shockwaves can play a role in an alternative and noninvasive treatment of biofilms by disrupting the biofilm and returning the bacteria to their planktonic form, making them much more vulnerable to antibiotic treatments.

Francis et al. [153] conducted an ex vivo study on pigskin, which, by scanning electron microscopy, showed that laser-generated shockwaves (LGS) were able to reduce biofilm coverage by 52% and significantly reduce the average cluster size. In another in vitro study by Yao et al. [154], biofilms were treated with gentamicin and LGS and a control group was treated with gentamicin alone. The LGS group showed that the biofilm burden, measured by the crystal violet biofilm assay, was significantly reduced and that the CFU density also significantly decreased. The application of LGS in tandem with antibiotic therapy may become an important treatment scheme because it is cost effective and its noninvasive nature provides distinct advantages.

References

1 Schmidt-Bleek K, Petersen A, Dienelt A, Schwarz C, Duda GN: Initiation and early control of tissue regeneration – bone healing as a model system for tissue regeneration. Expert Opin Biol Ther 2014;14: 247–259.

2 Giannoudis PV, Kontakis G: Treatment of long bone aseptic non-unions: Monotherapy or polytherapy? Injury 2009;40:1021–1022.

3 Kusnezov N, Prabhakar G, Dallo M, Thabet AM, Abdelgawad AA: Bone grafting via reamer-irrigator-aspirator for nonunion of open gustilo-anderson type iii tibial fractures treated with multiplanar external fixator. Sicot J 2017;3:30.

4 Masquelet AC, Begue T: The concept of induced membrane for reconstruction of long bone defects. Orthop Clin North Am 2010;41:27–37; table of contents.

5 Pederson WC, Person DW: Long bone reconstruction with vascularized bone grafts. Orthop Clin North Am 2007;38:23–35, v.

6 Giannoudis PV, Jones E, Einhorn TA: Fracture healing and bone repair. Injury 2011;42:549–550.

7 Morshed S, Corrales L, Genant H, Miclau T 3rd: Outcome assessment in clinical trials of fracture-healing. J Bone Joint Surg Am 2008;90(suppl 1):62–67.

8 Jager M, Hernigou P, Zilkens C, Herten M, Li X, Fischer J, Krauspe R: Cell therapy in bone healing disorders. Orthop Rev (Pavia) 2010;2:e20.

9 Dominici M, Le Blanc K, Mueller I, Slaper-Cortenbach I, Marini F, Krause D, Deans R, Keating A, Prockop D, Horwitz E: Minimal criteria for defining multipotent mesenchymal stromal cells. The international society for cellular therapy position statement. Cytotherapy 2006;8:315–317.

10 Crisan M, Yap S, Casteilla L, Chen CW, Corselli M, Park TS, Andriolo G, Sun B, Zheng B, Zhang L, Norotte C, Teng PN, Traas J, Schugar R, Deasy BM, Badylak S, Buhring HJ, Giacobino JP, Lazzari L, Huard J, Peault B: A perivascular origin for mesenchymal stem cells in multiple human organs. Cell Stem Cell 2008;3:301–313.

11 Corselli M, Chen CW, Sun B, Yap S, Rubin JP, Peault B: The tunica adventitia of human arteries and veins as a source of mesenchymal stem cells. Stem Cells Dev 2012;21:1299–1308.

12 Zuo C, Huang Y, Bajis R, Sahih M, Li YP, Dai K, Zhang X: Osteoblastogenesis regulation signals in bone remodeling. Osteoporos Int 2012;23:1653–1663.

13 Chen G, Deng C, Li YP: TGF-β and BMP signaling in osteoblast differentiation and bone formation. Int J Biol Sci 2012;8:272–288.

14 Caplan AI: New era of cell-based orthopedic therapies. Tissue Eng Part B Rev 2009;15:195–200.

15 Murray IR, Corselli M, Petrigliano FA, Soo C, Peault B: Recent insights into the identity of mesenchymal stem cells: Implications for orthopaedic applications. Bone Joint J 2014;96-B:291–298.

16 Jones BJ, McTaggart SJ: Immunosuppression by mesenchymal stromal cells: From culture to clinic. Exp Hematol 2008;36:733–741.

17 Haupt G: Use of extracorporeal shock waves in the treatment of pseudarthrosis, tendinopathy and other orthopedic diseases. J Urol 1997;158:4–11.

18 Wang CJ, Chen HS, Chen CE, Yang KD: Treatment of nonunions of long bone fractures with shock waves. Clin Orthop Relat Res 2001:95–101.

19 Chen YJ, Wurtz T, Wang CJ, Kuo YR, Yang KD, Huang HC, Wang FS: Recruitment of mesenchymal stem cells and expression of TGF-beta 1 and VEGF in the early stage of shock wave-promoted bone regeneration of segmental defect in rats. J Orthop Res 2004;22:526–534.

20 Wang FS, Wang CJ, Huang HJ, Chung H, Chen RF, Yang KD: Physical shock wave mediates membrane hyperpolarization and ras activation for osteogenesis in human bone marrow stromal cells. Biochem Biophys Res Commun 2001;287:648–655.

21 Wang FS, Yang KD, Chen RF, Wang CJ, Sheen-Chen SM: Extracorporeal shock wave promotes growth and differentiation of bone-marrow stromal cells towards osteoprogenitors associated with induction of TGF-beta1. J Bone Joint Surg Br 2002;84:457–461.

22 Schaden W, Fischer A, Sailler A: Extracorporeal shock wave therapy of nonunion or delayed osseous union. Clin Orthop Relat Res 2001;387:90–94.

23 Elster EA, Stojadinovic A, Forsberg J, Shawen S, Andersen RC, Schaden W: Extracorporeal shock wave therapy for nonunion of the tibia. J Orthop Trauma 2010;24:133–141.

24 Stojadinovic A, Kyle Potter B, Eberhardt J, Shawen SB, Andersen RC, Forsberg JA, Shwery C, Ester EA, Schaden W: Development of a prognostic naive bayesian classifier for successful treatment of nonunions. J Bone Joint Surg Am 2011;93:187–194.

25 Cacchio A, Giordano L, Colafarina O, Rompe JD, Tavernese E, Ioppolo F, Flamini S, Spacca G, Santilli V: Extracorporeal shock-wave therapy compared with surgery for hypertrophic long-bone nonunions. J Bone Joint Surg Am 2009;91:2589–2597.

26 Beutler S, Regel G, Pape HC, Machtens S, Weinberg AM, Kremeike I, Jonas U, Tscherne H: [extracorporeal shock wave therapy for delayed union of long bone fractures – preliminary results of a prospective cohort study]. Unfallchirurg 1999;102:839–847.

27 Xu ZH, Jiang Q, Chen DY, Xiong J, Shi DQ, Yuan T, Zhu XL: Extracorporeal shock wave treatment in nonunions of long bone fractures. Int Orthop 2009; 33:789–793.

28 Tawonsawatruk T, Kelly M, Simpson H: Evaluation of native mesenchymal stem cells from bone marrow and local tissue in an atrophic nonunion model. Tissue Eng Part C Methods 2014;20:524–532.

29 Stegen S, van Gastel N, Carmeliet G: Bringing new life to damaged bone: The importance of angiogenesis in bone repair and regeneration. Bone 2015;70: 19–27.

30 Wang CJ, Wang FS, Yang KD, Weng LH, Hsu CC, Huang CS, Yang LC: Shock wave therapy induces neovascularization at the tendon-bone junction. A study in rabbits. J Orthop Res 2003;21:984–989.

31 Wang CJ, Wang FS, Yang KD: Biological effects of extracorporeal shockwave in bone healing: a study in rabbits. Arch Orthop Trauma Surg 2008;128:879–884.

32 Sansone V, D' Agostino MC, Bonora C, Sizzano F, De Girolamo L, Romeo P: Early angiogenic response to shock waves in a three-dimensional model of human microvascular endothelial cell culture (HMEC-1). J Biol Regul Homeost Agents 2012;26:29–37.

33 Kuo SJ, Su IC, Wang CJ, Ko JY: Extracorporeal shockwave therapy (ESWT) in the treatment of atrophic non-unions of femoral shaft fractures. Int J Surg 2015;24:131–134.

34 Dumfarth J, Zimpfer D, Vogele-Kadletz M, Holfeld J, Sihorsch F, Schaden W, Czerny M, Aharinejad S, Wolner E, Grimm M: Prophylactic low-energy shock wave therapy improves wound healing after vein harvesting for coronary artery bypass graft surgery: a prospective, randomized trial. Ann Thorac Surg 2008;86:1909–1913.

35 Vigano M, Sansone V, D'Agostino MC, Romeo P, Perucca Orfei C, de Girolamo L: Mesenchymal stem cells as therapeutic target of biophysical stimulation for the treatment of musculoskeletal disorders. J Orthop Surg Res 2016;11:163.

36 Suhr F, Delhasse Y, Bungartz G, Schmidt A, Pfannkuche K, Bloch W: Cell biological effects of mechanical stimulations generated by focused extracorporeal shock wave applications on cultured human bone marrow stromal cells. Stem Cell Res 2013; 11:951–964.

37 BrañEs J, Contreras HR, Cabello P, Antonic V, Guiloff LJ, Brañes M: Shoulder rotator cuff responses to extracorporeal shockwave therapy: morphological and immunohistochemical analysis. Should Elb 2012;4:163–168.

38 Hofmann A, Ritz U, Hessmann MH, Alini M, Rommens PM, Rompe JD: Extracorporeal shock wave-mediated changes in proliferation, differentiation, and gene expression of human osteoblasts. J Trauma 2008;65:1402–1410.

39 Wang FS, Yang KD, Kuo YR, Wang CJ, Sheen-Chen SM, Huang HC, Chen YJ: Temporal and spatial expression of bone morphogenetic proteins in extracorporeal shock wave-promoted healing of segmental defect. Bone 2003;32:387–396.

40 Tam KF, Cheung WH, Lee KM, Qin L, Leung KS: Shockwave exerts osteogenic effect on osteoporotic bone in an ovariectomized goat model. Ultrasound Med Biol 2009;35:1109–1118.

41 Catalano MG, Marano F, Rinella L, de Girolamo L, Bosco O, Fortunati N, Berta L, Frairia R: Extracorporeal shockwaves (ESWS) enhance the osteogenic medium-induced differentiation of adipose-derived stem cells into osteoblast-like cells. J Tissue Eng Regen Med 2017;11:390–399.

42 Baxter MA, Wynn RF, Jowitt SN, Wraith JE, Fairbairn LJ, Bellantuono I: Study of telomere length reveals rapid aging of human marrow stromal cells following in vitro expansion. Stem Cells 2004;22: 675–682.

43 Strand OT, Goosman DR, Martinez C, Whitworth TL, Kuhlow WW: Compact system for high-speed velocimetry using heterodyne techniques. Rev Sci Inst 2006;77:083108.

44 Delius M: Medical applications and bioeffects of extracorporeal shock waves. Shock Waves 1994;4:55–72.

45 Brümmer F, Bräuner T, Hülser DF: Biological effects of shock waves. World J Urol 1990;8:224–232.

46 Kohri K, Uemura T, Iguchi M, Kurita T: Effect of high energy shock waves on tumor cells. Urol Res 1990;18:101–105.

47 Clayman RV, Long S, Marcus M: High-energy shock waves: in vitro effects. Am J Kidney Dis 1991;17: 436–444.

48 Delius M, Denk R, Berding C, Liebich HG, Jordan M, Brendel W: Biological effects of shock waves: cavitation by shock waves in piglet liver. Ultrasound Med Biol 1990;16:467–472.

49 Huber P, Debus J, Peschke P, Hahn EW, Lorenz WJ: In vivo detection of ultrasonically induced cavitation by a fibre-optic technique. Ultrasound Med Biol 1994;20:811–825.

50 Cleveland RO, Sapozhnikov OA, Bailey MR, Crum LA: A dual passive cavitation detector for localized detection of lithotripsy-induced cavitation in vitro. J Acoust Soc Am 2000;107:1745–1758.

51 Delius M: Extracorporeal shock waves: Bioeffects and mechanisms of action; in Srivastava RC, Leutloff D, Takayama K, Grönig H (eds): Shock Focusing Effect in Medical Science and Sonoluminescence. Berlin/Heidelberg, Springer, 2003, pp 211–266.

52 Russo P, Stephenson RA, Mies C, Huryk R, Heston WD, Melamed MR, Fair WR: High energy shock waves suppress tumor growth in vitro and in vivo. J Urol 1986;135:626–628.

53 Worle K, Steinbach P, Hofstadter F: The combined effects of high-energy shock waves and cytostatic drugs or cytokines on human bladder cancer cells. Br J Cancer 1994;69:58–65.

54 Randazzo RF, Chaussy CG, Fuchs GJ, Bhuta SM, Lovrekovich H, deKernion JB: The in vitro and in vivo effects of extracorporeal shock waves on malignant cells. Urol Res 1988;16:419–426.

55 Delius M, Adams G: Shock wave permeabilization with ribosome inactivating proteins: a new approach to tumor therapy. Cancer Res 1999;59:5227–5232.

56 Kodama T, Hamblin MR, Doukas AG: Cytoplasmic molecular delivery with shock waves: importance of impulse. Biophys J 2000;79:1821–1832.

57 Kodama T, Doukas AG, Hamblin MR: Delivery of ribosome-inactivating protein toxin into cancer cells with shock waves. Cancer Lett 2003;189:69–75.

58 Michel MS, Erben P, Trojan L, Schaaf A, Kiknavelidze K, Knoll T, Alken P: Acoustic energy: a new transfection method for cancer of the prostate, cancer of the bladder and benign kidney cells. Anticancer Res 2004;24:2303–2308.

59 Ohl CD, Arora M, Ikink R, de Jong N, Versluis M, Delius M, Lohse D: Sonoporation from jetting cavitation bubbles. Biophys J 2006;91:4285–4295.

60 Moosavi-Nejad SF, Hosseini SH, Satoh M, Takayama K: Shock wave induced cytoskeletal and morphological deformations in a human renal carcinoma cell line. Cancer Sci 2006;97:296–304.

61 Oosterhof GO, Smits GA, de Ruyter AE, Schalken JA, Debruyne FM: Effects of high-energy shock waves combined with biological response modifiers in different human kidney cancer xenografts. Ultrasound Med Biol 1991;17:391–399.

62 Chen CL, Guo ZH, Zhao Y, Yan CY, Pu JX, Chen ZX, Zhang R: Suppressive effect of high energy shock waves on tumor cells. Chin Med J (Engl) 1991;104:548–551.

63 Gamarra F, Spelsberg F, Dellian M, Goetz AE: Complete local tumor remission after therapy with extracorporeally applied high-energy shock waves (HESW). Int J Cancer 1993;55:153–156.

64 Frairia R, Catalano MG, Fortunati N, Fazzari A, Raineri M, Berta L: High energy shock waves (hesw) enhance paclitaxel cytotoxicity in mcf-7 cells. Breast Cancer Res Treat 2003;81:11–19.

65 Millan-Chiu B, Camacho G, Varela-Echavarria A, Tamariz E, Fernandez F, Lopez-Marin LM, Loske AM: Shock waves and DNA-cationic lipid assemblies: a synergistic approach to express exogenous genes in human cells. Ultrasound Med Biol 2014;40:1599–1608.

66 Coleman AJ, Saunders JE: A review of the physical properties and biological effects of the high amplitude acoustic field used in extracorporeal lithotripsy. Ultrasonics 1993;31:75–89.

67 Huber PE, Debus J: Tumor cytotoxicity in vivo and radical formation in vitro depend on the shock wave-induced cavitation dose. Radiat Res 2001;156:301–309.

68 Frairia R, Berta L, Catalano MG: Extracorporeal shock waves: perspectives in malignant tumor treatment. J Biol Regul Homeost Agents 2016;30:641–648.

69 Brummer F, Suhr D, Hulser DF: Sensitivity of normal and malignant cells to shock waves. J Stone Dis 1992;4:243–248.

70 van Dongen JW, van Steenbrugge GJ, Romijn JC, Schroder FH: The cytocidal effect of high energy shock waves on human prostatic tumour cell lines. Eur J Cancer Clin Oncol 1989;25:1173–1179.

71 Hochstrasser T, Frank HG, Schmitz C: Dose-dependent and cell type-specific cell death and proliferation following in vitro exposure to radial extracorporeal shock waves. Sci Rep 2016;6:30637.

72 Brummer F, Brenner J, Brauner T, Hulser DF: Effect of shock waves on suspended and immobilized l1210 cells. Ultrasound Med Biol 1989;15:229–239.

73 Kodama T, Doukas AG, Hamblin MR: Shock wave-mediated molecular delivery into cells. Biochim Biophys Acta 2002;1542:186–194.

74 Gambihler S, Delius M: In vitro interaction of lithotripter shock waves and cytotoxic drugs. Br J Cancer 1992;66:69–73.

75 Kato M, Ioritani N, Suzuki T, Kambe M, Inaba Y, Watanabe R, Sasano H, Orikasa S: Mechanism of anti-tumor effect of combination of bleomycin and shock waves. Jpn J Cancer Res 2000;91:1065–1072.

76 Qi B, Yu T, Wang C, Wang T, Yao J, Zhang X, Deng P, Xia Y, Junger WG, Sun D: Shock wave-induced atp release from osteosarcoma U2OS cells promotes cellular uptake and cytotoxicity of methotrexate. J Exp Clin Cancer Res 2016;35:161.

77 McAuliffe DJ, Lee S, Flotte TJ, Doukas AG: Stress-wave-assisted transport through the plasma membrane in vitro. Lasers Surg Med 1997;20:216–222.

78 Mulholland SE, Lee S, McAuliffe DJ, Doukas AG: Cell loading with laser-generated stress waves: the role of the stress gradient. Pharm Res 1999;16:514–518.

79 Hofmann GA, Dev SB, Dimmer S, Nanda GS: Electroporation therapy: A new approach for the treatment of head and neck cancer. IEEE Trans Biomed Eng 1999;46:752–759.

80 Zimmermann KC, Bonzon C, Green DR: The machinery of programmed cell death. Pharmacol Ther 2001;92:57–70.

81 Steller H: Mechanisms and genes of cellular suicide. Science 1995;267:1445–1449.

82 Chen X, Kolb JF, Swanson RJ, Schoenbach KH, Beebe SJ: Apoptosis initiation and angiogenesis inhibition: melanoma targets for nanosecond pulsed electric fields. Pigment Cell Melanoma Res 2010;23: 554–563.

83 Stacey M, Fox P, Buescher S, Kolb J: Nanosecond pulsed electric field induced cytoskeleton, nuclear membrane and telomere damage adversely impact cell survival. Bioelectrochemistry 2011;82:131–134.

84 Ashush H, Rozenszajn LA, Blass M, Barda-Saad M, Azimov D, Radnay J, Zipori D, Rosenschein U: Apoptosis induction of human myeloid leukemic cells by ultrasound exposure. Cancer Res 2000;60: 1014.

85 Bayrak O, Cimentepe E, Karatas OF, Aker A, Bayrak R, Yildirim ME, Unsal A, Unal D: Ovarian apoptosis after shock wave lithotripsy for distal ureteral stones. Urol Res 2009;37:69–74.

86 Gecit I, Kavak S, Oguz EK, Pirincci N, Gunes M, Kara M, Ceylan K, Kaba M, Tanik S: Tissue damage in kidney, adrenal glands and diaphragm following extracorporeal shock wave lithotripsy. Toxicol Ind Health 2014;30:845–850.

87 Lukes P, Zeman J, Horak V, Hoffer P, Pouckova P, Holubova M, Hosseini SH, Akiyama H, Sunka P, Benes J: In vivo effects of focused shock waves on tumor tissue visualized by fluorescence staining techniques. Bioelectrochemistry 2015;103:103–110.

88 Palmero A, Berger M, Venturi C, Ferrero I, Rustichelli D, Berta L, Frairia R, Madon E, Fagioli F: High energy shock waves enhance the cytotoxic effect of doxorubicin and methotrexate to human osteosarcoma cell lines. Oncol Rep 2006;15:267–273.

89 Canaparo R, Serpe L, Catalano MG, Bosco O, Zara GP, Berta L, Frairia R: High energy shock waves (hesw) for sonodynamic therapy: effects on ht-29 human colon cancer cells. Anticancer Res 2006;26: 3337–3342.

90 Catalano MG, Costantino L, Fortunati N, Bosco O, Pugliese M, Boccuzzi G, Berta L, Frairia R: High energy shock waves activate 5'-aminolevulinic acid and increase permeability to paclitaxel: antitumor effects of a new combined treatment on anaplastic thyroid cancer cells. Thyroid 2007;17:91–99.

91 Fukuda H, Casas A, Batlle A: Aminolevulinic acid: from its unique biological function to its star role in photodynamic therapy. Int J Biochem Cell Biol 2005; 37:272–276.

92 Burch S, Bogaards A, Siewerdsen J, Moseley D, Yee A, Finkelstein J, Weersink R, Wilson BC, Bisland SK: Photodynamic therapy for the treatment of metastatic lesions in bone: studies in rat and porcine models. J Biomed Opt 2005;10:034011.

93 Peng Q, Warloe T, Berg K, Moan J, Kongshaug M, Giercksky KE, Nesland JM: 5-Aminolevulinic acid-based photodynamic therapy. Clinical research and future challenges. Cancer 1997;79:2282–2308.

94 Agostinis P, Berg K, Cengel KA, Foster TH, Girotti AW, Gollnick SO, Hahn SM, Hamblin MR, Juzeniene A, Kessel D, Korbelik M, Moan J, Mroz P, Nowis D, Piette J, Wilson BC, Golab J: Photodynamic therapy of cancer: an update. CA Cancer J Clin 2011;61:250–281.

95 Yumita N, Okuyama N, Sasaki K, Umemura S: Sonodynamic therapy on chemically induced mammary tumor: pharmacokinetics, tissue distribution and sonodynamically induced antitumor effect of porfimer sodium. Cancer Sci 2004;95:765–769.

96 Kuroki M, Hachimine K, Abe H, Shibaguchi H, Maekawa S, Yanagisawa J, Kinugasa T, Tanaka T, Yamashita Y: Sonodynamic therapy of cancer using novel sonosensitizers. Anticancer Res 2007;27: 3673–3677.

97 Tachibana K, Feril LB Jr, Ikeda-Dantsuji Y: Sonodynamic therapy. Ultrasonics 2008;48:253–259.

98 Ashokkumar M: The characterization of acoustic cavitation bubbles – an overview. Ultrason Sonochem 2011;18:864–872.

99 Rosenthal I, Sostaric JZ, Riesz P: Sonodynamic therapy – a review of the synergistic effects of drugs and ultrasound. Ultrason Sonochem 2004;11:349–363.

100 Serpe L, Canaparo R, Berta L, Bargoni A, Zara GP, Frairia R: High energy shock waves and 5-aminolevulinic for sonodynamic therapy: effects in a syngeneic model of colon cancer. Technol Cancer Res Treat 2011;10:85–93.

101 Canaparo R, Serpe L, Zara GP, Chiarle R, Berta L, Frairia R: High energy shock waves (HESW) increase paclitaxel efficacy in a syngeneic model of breast cancer. Technol Cancer Res Treat 2008;7: 117–124.

102 Foglietta F, Canaparo R, Francovich A, Arena F, Civera S, Cravotto G, Frairia R, Serpe L: Sonodynamic treatment as an innovative bimodal anticancer approach: shock wave-mediated tumor growth inhibition in a syngeneic breast cancer model. Discov Med 2015;20:197–205.

103 Foglietta F, Duchi S, Canaparo R, Varchi G, Lucarelli E, Dozza B, Serpe L: Selective sensitiveness of mesenchymal stem cells to shock waves leads to anticancer effect in human cancer cell co-cultures. Life Sci 2017;173:28–35.

104 Krasovitski B, Frenkel V, Shoham S, Kimmel E: Intramembrane cavitation as a unifying mechanism for ultrasound-induced bioeffects. Proc Natl Acad Sci 2011;108:3258–3263.

105 Cao Y, Wang B, Lou D, Wang Y, Hao S, Zhang L: Nanoscale delivery systems for multiple drug combinations in cancer. Future Oncol 2011;7:1347–1357.

106 Yu X, Pishko MV: Nanoparticle-based biocompatible and targeted drug delivery: characterization and in vitro studies. Biomacromolecules 2011;12: 3205–3212.

107 Ahmed SE, Martins AM, Husseini GA: The use of ultrasound to release chemotherapeutic drugs from micelles and liposomes. J Drug Target 2015;23:16–42.

108 Michalet X, Pinaud FF, Bentolila LA, Tsay JM, Doose S, Li JJ, Sundaresan G, Wu AM, Gambhir SS, Weiss S: Quantum dots for live cells, in vivo imaging, and diagnostics. Science 2005;307:538–544.

109 Kaestner P, Aigner A, Bastians H: Therapeutic targeting of the mitotic spindle checkpoint through nanoparticle-mediated sirna delivery inhibits tumor growth in vivo. Cancer Lett 2011;304:128–136.

110 Maeda H: Tumor-selective delivery of macromolecular drugs via the epr effect: background and future prospects. Bioconjug Chem 2010;21:797–802.

111 Davis ME, Chen ZG, Shin DM: Nanoparticle therapeutics: an emerging treatment modality for cancer. Nat Rev Drug Discov 2008;7:771–782.

112 Slingerland M, Guchelaar HJ, Gelderblom H: Liposomal drug formulations in cancer therapy: 15 years along the road. Drug Discov Today 2012;17:160–166.

113 Ishida O, Maruyama K, Sasaki K, Iwatsuru M: Size-dependent extravasation and interstitial localization of polyethyleneglycol liposomes in solid tumor-bearing mice. Int J Pharm 1999;190:49–56.

114 Ganta S, Devalapally H, Shahiwala A, Amiji M: A review of stimuli-responsive nanocarriers for drug and gene delivery. J Control Release 2008;126:187–204.

115 Rapoport A, Marin A, Christensen D: Ultrasound-activated drug delivery. Drug Deliv Syst Sci 2002:37–46.

116 Sazgarnia A, Shanei A, Meibodi NT, Eshghi H, Nassirli H: A novel nanosonosensitizer for sonodynamic therapy: in vivo study on a colon tumor model. J Ultrasound Med 2011;30:1321–1329.

117 Marano F, Argenziano M, Frairia R, Adamini A, Bosco O, Rinella L, Fortunati N, Cavalli R, Catalano MG: Doxorubicin-loaded nanobubbles combined with extracorporeal shock waves: basis for a new drug delivery tool in anaplastic thyroid cancer. Thyroid 2016;26:705–716.

118 Tacar O, Sriamornsak P, Dass CR: Doxorubicin: An update on anticancer molecular action, toxicity and novel drug delivery systems. J Pharm Pharmacol 2013;65:157–170.

119 Marano F, Rinella L, Argenziano M, Cavalli R, Sassi F, D'Amelio P, Battaglia A, Gontero P, Bosco O, Peluso R, Fortunati N, Frairia R, Catalano MG: Targeting taxanes to castration-resistant prostate cancer cells by nanobubbles and extracorporeal shock waves. PLoS One 2016;11:e0168553.

120 Zhong J, Yang S, Wen L, Xing D: Imaging-guided photoacoustic drug release and synergistic chemo-photoacoustic therapy with paclitaxel-containing nanoparticles. J Control Release 2016;226:77–87.

121 Serpe L, Foglietta F, Canaparo R: Nanosonotechnology: the next challenge in cancer sonodynamic therapy. Nanotechnol Rev 2012;1:173.

122 Canaparo R, Varchi G, Ballestri M, Foglietta F, Sotgiu G, Guerrini A, Francovich A, Civera P, Frairia R, Serpe L: Polymeric nanoparticles enhance the sonodynamic activity of meso-tetrakis (4-sulfonatophenyl) porphyrin in an in vitro neuroblastoma model. Int J Nanomed 2013;8:4247–4263.

123 Ginn SL, Alexander IE, Edelstein ML, Abedi MR, Wixon J: Gene therapy clinical trials worldwide to 2012 – an update. J Gene Med 2013;15:65–77.

124 Pahle J, Walther W: Vectors and strategies for non-viral cancer gene therapy. Expert Opin Biol Ther 2016;16:443–461.

125 Ha CH, Lee SC, Kim S, Chung J, Bae H, Kwon K: Novel mechanism of gene transfection by low-energy shock wave. Sci Rep 2015;5:12843.

126 Mehier-Humbert S, Guy RH: Physical methods for gene transfer: improving the kinetics of gene delivery into cells. Adv Drug Deliv Rev 2005;57:733–753.

127 Chong H, Todryk S, Hutchinson G, Hart IR, Vile RG: Tumour cell expression of b7 costimulatory molecules and interleukin-12 or granulocyte-macrophage colony-stimulating factor induces a local antitumour response and may generate systemic protective immunity. Gene Therapy 1998;5:223–232.

128 Lasek W, Zagozdzon R, Jakobisiak M: Interleukin 12: still a promising candidate for tumor immunotherapy? Cancer Immunol Immunother 2014;63:419–435.

129 Miller DL, Song J: Lithotripter shock waves with cavitation nucleation agents produce tumor growth reduction and gene transfer in vivo. Ultrasound Med Biol 2002;28:1343–1348.

130 Denies S, Cicchelero L, Van Audenhove I, Sanders NN: Combination of interleukin-12 gene therapy, metronomic cyclophosphamide and DNA cancer vaccination directs all arms of the immune system towards tumor eradication. J Control Release 2014;187:175–182.

131 Rodriguez-Madoz JR, Zabala M, Alfaro M, Prieto J, Kramer MG, Smerdou C: Short-term intratumoral interleukin-12 expressed from an alphaviral vector is sufficient to induce an efficient antitumoral response against spontaneous hepatocellular carcinomas. Hum Gene Ther 2014;25:132–143.

132 Berger M, Frairia R, Piacibello W, Sanavio F, Palmero A, Venturi C, Pignochino Y, Berta L, Madon E, Aglietta M, Fagioli F: Feasibility of cord blood stem cell manipulation with high-energy shock waves: an in vitro and in vivo study. Exp Hematol 2005;33:1371–1387.

133 Murata R, Nakagawa K, Ohtori S, Ochiai N, Arai M, Saisu T, Sasho T, Takahashi K, Moriya H: The effects of radial shock waves on gene transfer in rabbit chondrocytes in vitro. Osteoarthritis Cartilage 2007;15:1275–1282.

134 Fan Z, Chen D, Deng CX: Improving ultrasound gene transfection efficiency by controlling ultrasound excitation of microbubbles. J Control Release 2013;170:401–413.

135 Yoon YI, Kwon YS, Cho HS, Heo SH, Park KS, Park SG, Lee SH, Hwang SI, Kim YI, Jae HJ, Ahn GJ, Cho YS, Lee H, Lee HJ, Yoon TJ: Ultrasound-mediated gene and drug delivery using a microbubble-liposome particle system. Theranostics 2014;4:1133–1144.

136 Zhou Q-L, Chen Z-Y, Wang Y-X, Yang F, Lin Y, Liao Y-Y: Ultrasound-mediated local drug and gene delivery using nanocarriers. Biomed Res Int 2014;2014:13.

137 Jessen KR, Mirsky R: The origin and development of glial cells in peripheral nerves. Nat Rev Neurosci 2005;6:671–682.

138 Zhou Y, Notterpek L: Promoting peripheral myelin repair. Exp Neurol 2016;283:573–580.

139 Maggi SP, Lowe JB, 3rd, Mackinnon SE: Pathophysiology of nerve injury. Clin Plast Surg 2003;30:109–126.

140 Isaacs J: Treatment of acute peripheral nerve injuries: current concepts. J Hand Surg Am 2010;35:491–497; quiz 498.

141 Hausner T, Pajer K, Halat G, Hopf R, Schmidhammer R, Redl H, Nogradi A: Improved rate of peripheral nerve regeneration induced by extracorporeal shock wave treatment in the rat. Exp Neurol 2012;236:363–370.

142 Schuh CM, Hercher D, Stainer M, Hopf R, Teuschl AH, Schmidhammer R, Redl H: Extracorporeal shockwave treatment: a novel tool to improve Schwann cell isolation and culture. Cytotherapy 2016;18:760–770.

143 Yamaya S, Ozawa H, Kanno H, Kishimoto KN, Sekiguchi A, Tateda S, Yahata K, Ito K, Shimokawa H, Itoi E: Low-energy extracorporeal shock wave therapy promotes vascular endothelial growth factor expression and improves locomotor recovery after spinal cord injury. J Neurosurg 2014;121:1514–1525.

144 Yahata K, Kanno H, Ozawa H, Yamaya S, Tateda S, Ito K, Shimokawa H, Itoi E: Low-energy extracorporeal shock wave therapy for promotion of vascular endothelial growth factor expression and angiogenesis and improvement of locomotor and sensory functions after spinal cord injury. J Neurosurg Spine 2016;25:745–755.

145 Lee JH, Kim SG: Effects of extracorporeal shock wave therapy on functional recovery and neurotrophin-3 expression in the spinal cord after crushed sciatic nerve injury in rats. Ultrasound Med Biol 2015;41:790–796.

146 Rosenstein JM, Krum JM: New roles for vegf in nervous tissue – beyond blood vessels. Exp Neurol 2004;187:246–253.

147 Long HQ, Li GS, Cheng X, Xu JH, Li FB: Role of hypoxia-induced vegf in blood-spinal cord barrier disruption in chronic spinal cord injury. Chin J Traumatol 2015;18:293–295.

148 Nagoshi N, Okano H: Applications of induced pluripotent stem cell technologies in spinal cord injury. J Neurochem 2017;141:848–860.

149 Lee JY, Ha KY, Kim JW, Seo JY, Kim YH: Does extracorporeal shock wave introduce alteration of microenvironment in cell therapy for chronic spinal cord injury? Spine (Phila Pa 1976) 2014;39:E1553–E1559.

150 Lee JH, Cho SH: Effect of extracorporeal shock wave therapy on denervation atrophy and function caused by sciatic nerve injury. J Phys Ther Sci 2013;25:1067–1069.

151 Nakagomi T, Nakano-Doi A, Kawamura M, Matsuyama T: Do vascular pericytes contribute to neurovasculogenesis in the central nervous system as multipotent vascular stem cells? Stem Cells Dev 2015;24:1730–1739.

152 Kizhner V, Krespi YP, Hall-Stoodley L, Stoodley P: Laser-generated shockwave for clearing medical device biofilms. Photomed Laser Surg 2011;29:277–282.

153 Francis NC, Yao W, Grundfest WS, Taylor ZD: Laser-generated shockwaves as a treatment to reduce bacterial load and disrupt biofilm. IEEE Trans Biomed Eng 2017;64:882–889.

154 Yao W, Kuan EC, Francis NC, St John MA, Grundfest WS, Taylor ZD: Laser-generated shockwaves enhance antibacterial activity against biofilms in vitro. Lasers Surg Med 2017;49:539–547.

Valerio Sansone, MD
Istituto Ortopedico Galeazzi IRCCS
Via Riccardo Galeazzi 4
IT–20161 Milan (Italy)
E-Mail valerio.sansone@unimi.it

Author Index

Subject Index